REVISE OCR AS/A LEVEL
Chemistry

D0496374

LEARNING R

REVISION GUIDE

Series Consultant: Harry Smith

Authors: Mark Grinsell and David Brentnall

Our revision resources are the smart choice for those revising for OCR AS/A Level Chemistry. This book will help you to:

- **Organise** your revision with the one-topic-per-page format

- **Speed up** your revision with summary notes in short, memorable chunks

- **Track** your revision progress with at-a-glance check boxes

- **Check** your understanding with worked examples

- **Develop** your exam technique with exam-style practice questions and full answers.

Revision is more than just this Guide!

Make sure that you have practised every topic covered in this book, with the accompanying OCR AS/A Level Chemistry Revision Workbook. It gives you:

- More exam-style practice and a 1-to-1 page match with this Revision Guide

- Guided questions to help build your confidence

- Hints to support your revision and practice.

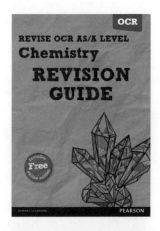

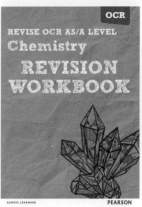

For the full range of Pearson revision titles across GCSE, BTEC and AS/A Level visit:

www.pearsonschools.co.uk/revise

ALWAYS LEARNING

PEARSON

Contents

1-to-1 page match with the Chemistry Revision Workbook ISBN 9781447984320

Atomic structure and isotopes

The model of the atom describes atoms or ions in terms of particles – protons, neutrons and electrons.

Properties of the particles found in atoms

The current model of the atom consists of protons and neutrons in a dense nucleus and electrons around the nucleus in 'shells'. The mass of the atom is concentrated in the nucleus but it takes up only a tiny fraction of the space. The relative masses and charges of the atomic particles are shown in the table:

Particle	Relative mass	Relative charge
proton	1.0	1+
neutron	1.0	0
electron	$\frac{1}{2000}$	1−

The Periodic Table

Information from the Periodic Table can be used to deduce the numbers of protons, neutrons and electrons found in an atom. Look at potassium:

Mass number: the total number of protons AND neutrons in the nucleus.

$$^{39}_{19}\text{K} \quad \text{Potassium}$$

Atomic number: the number of protons in the nucleus. For a neutral atom this is equivalent to the number of electrons orbiting in shells.

$$\text{number of neutrons} = \text{mass number} - \text{atomic number}$$

For potassium, number of neutrons = 39 − 19

= 20

Isotopes

Isotopes are atoms of the same element with different numbers of neutrons and hence different mass numbers. They have the same atomic number so the number of protons is the same.

Some isotopes occur naturally, such as carbon isotopes, others are formed artificially in nuclear reactions. Some isotopes of heavier elements are unstable and emit radiation, they are radioactive isotopes.

Be careful! When defining the word isotope be precise. Isotopes are **atoms of the same element** with a different number of neutrons.

Examples of isotopes of carbon

Isotope	$^{12}_{6}\text{C}$	$^{13}_{6}\text{C}$	$^{14}_{6}\text{C}$
Mass number	12	13	14
Atomic number	6	6	6
Number of neutrons	6	7	8

Although the mass numbers of the different isotopes of carbon are different, they are chemically identical. This is because:

- chemical reactions involve only the electrons
- each isotope has an identical number of electrons.

Worked example

Deduce the numbers of protons, neutrons and electrons in the 2+ ion formed by the isotope Mg-24

Protons = 12; Neutrons = 12; Electrons = 10

From the Periodic Table the atomic number of magnesium is 12.

Atomic number is still equal to the number of protons as they don't change when an ion forms.

24 is the mass number and the total number of protons and neutrons in the nucleus. So, the number of neutrons will be 24 − 12 = 12 neutrons.

The question concerns the Mg^{2+} ion. The number of electrons will be reduced by 2 so there will be 10 electrons not 12.

Now try this

Write the symbols of the particles containing:

29 protons, 29 electrons and 35 neutrons

11 protons, 10 electrons and 12 neutrons

17 protons, 18 electrons and 18 neutrons

8 protons, 7 electrons and 8 neutrons.   **(4 marks)**

Relative masses

As atoms are so small chemists use the idea of relative masses to compare the masses of atoms, elements and compounds.

Relative isotopic mass

Relative isotopic mass is defined as:

mass of an atom of the isotope compared with $\frac{1}{12}$th of the mass of an atom of carbon-12.

Remember that isotopes of the same element will have different isotopic masses, such as ^{35}Cl and ^{37}Cl.

Relative atomic mass

It is unlikely that all atoms in a sample will contain only one isotope, so relative atomic mass is often more useful. It is defined as:

weighted mean mass of all isotopes of the element (in a sample) compared with $\frac{1}{12}$th of the mass of an atom of carbon-12.

Relative molecular mass, M_r

This term is used when referring to molecules with simple covalent structures and is the sum of the atomic masses of all the atoms present in the substance.

Relative formula mass

This term is used when referring to substances with giant structures and is the sum of the atomic masses of all the atoms in the formula of the substance.

> You need to use the correct terminology to describe the relative mass for a given substance and to define relative isotopic and relative atomic masses in relation to the mass of a carbon-12 atom.

Worked example

Calculate the relative molecular mass of ethanol, C_2H_5OH. **(1 mark)**

Ethanol contains 2 C atoms	$= 2 \times 12.0$	$= 24.0$
6 H atoms	$= 6 \times 1.0$	$= 6.0$
and 1 O atom	$= 1 \times 16.0$	$= 16.0$
so relative molecular mass		$= \overline{46.0}$

> You don't have to include all the stages of working out shown here but you may find it helpful when practising such examples, especially with more complex compounds.

Worked example

Calculate the relative formula mass of hydrated copper sulfate, $CuSO_4 \cdot 5H_2O$. **(1 mark)**

Hydrated copper sulfate contains

1 Cu atom	$= 1 \times 63.5 =$	63.5
1 S atom	$= 1 \times 32.1 =$	32.1
4 O atoms	$= 4 \times 16.0 =$	64.0
and 5 H_2O molecules	$= 5 \times 18.0 =$	90.0
so relative formula mass	$=$	$\overline{249.6}$

> The Periodic Table used at A level contains more precise values for atomic masses. When an element has more than one common isotope some of these values are not whole numbers. For instance, 32 is often recalled for sulfur, but its precise value is 32.1 as used here.

> Remember this includes the water molecules as well, which are included in the relative formula mass of hydrated salts.

Now try this

Calculate the relative formula mass of Mohr's salt, $(NH_4)_2Fe(SO_4)_2.6H_2O$.   **(1 mark)**

Using mass spectroscopy

Mass spectrometers can be used to find the abundance of different isotopes present in a sample of an element. The results obtained can be used to calculate the relative atomic mass of a sample.

Worked example

A sample of the element zirconium contains five isotopes. The percentage of each is shown in the table.

Isotope	Percentage in sample / %
Zr-90	53.1
Zr-91	10.9
Zr-92	16.2
Zr-94	16.7
Zr-96	3.1

Calculate the relative atomic mass of the sample and give your answer to one decimal place. **(3 marks)**

To find the answer you have to calculate a *weighted mean*. That's a mean value for all the isotopes, taking into account the percentage of each. In this case the answer will be

$$\frac{(90 \times 53.1) + (91 \times 10.9) + (92 \times 16.2) + (94 \times 16.7) + (96 \times 3.1)}{100} = 91.3$$

Maths skills As the question has specifically asked for an answer to one decimal place, the final mark may have been lost if the 'calculator' value (91.287), had been quoted.

Worked example

Use the mass spectrum to calculate the relative atomic mass of a sample of magnesium.

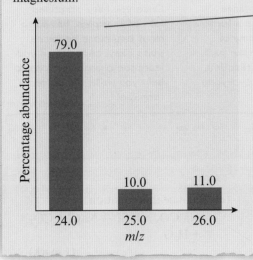

This is an alternative way the data may be presented but the method of calculation is the same. As the atoms are ionised in the mass spectrometer, the x axis, $\frac{m}{z}$, is the mass to charge ratio of the ions formed. However, this is equivalent to the mass of the atoms as only 1 electron is removed.

You have to find the percentage of each isotope from the mass spectrum, then calculate the weighted mean.

$$\frac{(24.0 \times 79.0) + (25.0 \times 10.0) + (26.0 \times 11.0)}{100} = 24.3$$

Note that as all data in the question is given to three significant figures, the final answer is also given to three significant figures.

Now try this

A sample of the element gallium consists of two isotopes, Ga-69 and Ga-71. The relative atomic mass of the sample was found to be 69.7. Calculate the percentage abundance of each isotope. **(3 marks)**

Use algebra to help solve the problem. Let x be the percentage abundance of Ga-69, so the percentage abundance of Ga-71 = (100 − x).

Writing formulae and equations

Chemists use formulae and equations to summarise the changes that occur in reactions. These are key skills that you need to practise in order to make progress in all branches of the subject.

Formulae

There are some ions you are expected to remember to help you write formulae. These are:

Ion	Formula
nitrate	NO_3^-
carbonate	CO_3^{2-}
sulfate	SO_4^{2-}
hydroxide	OH^-
ammonium	NH_4^+
zinc	Zn^{2+}
silver	Ag^+

Learn these by covering, recalling them, then uncovering to check until you are consistently getting them correct.

Predicting charges on ions

You should be able to predict the charge on some ions based on their position in the Periodic Table by following some simple rules, for example:

- Metal atoms in Groups 1, 2 and 3 form 1+, 2+ and 3+ ions respectively.
- Non-metal atoms in Groups 5, 6 and 7 can form 3^-, 2^- and 1^- ions respectively.

Using the charges on ions to write a formula

You can deduce the formula of a compound from its name provided you have learnt the formula and charges of the ions involved.

For example, here's how to work out the formula of calcium nitrate:

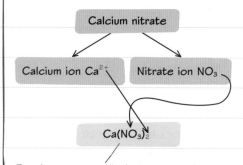

Brackets are used whenever you have more than one ion that consists of two or more types of atom, in this case, NO_3^-

Writing a balanced equation

You cannot simply create or destroy atoms in chemical reactions so the numbers of each atom on both sides of an equation have to balance, even if you rearranged the atoms into new substances. Look at the equation for the combustion of methane.

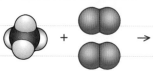

One methane molecule	Two oxygen molecules		One carbon dioxide molecule	Two water molecules
CH_4	+	$2O_2$	$\rightarrow$ CO_2	+ $2H_2O$
1C (4H)		(4O)	1C (2O)	2O (4H)

Remember the large numbers before a formula tell you how many of that molecule take part or form in the reaction, in this case two oxygen molecules.

Once you have practised lots of examples, balancing most equations becomes quite easy, especially if you learn the charges on ions to help you write the correct formulae!

Worked example

Write the formulae of
(a) sodium sulfate   **(1 mark)**

Na_2SO_4, two Na^+ ions needed to balance charge on SO_4^{2-}

(b) magnesium nitrate   **(1 mark)**

$Mg(NO_3)_2$, two NO_3^- ions needed to balance charge on Mg^{2+}

(c) Balance this equation
$$Na_2CO_3 + HCl \rightarrow NaCl + CO_2 + H_2O$$
                                                        (1 mark)

$$Na_2CO_3 + 2HCl \rightarrow 2NaCl + CO_2 + H_2O$$

Now try this

Write a balanced equation for the reaction between potassium hydroxide and sulfuric acid.   **(2 marks)**

Amount of substance – the mole

Chemists measure and calculate the amount of a substance using a unit called the mole.

The amount of substance

The amount of a substance is measured in *moles (mol)*. One mole of any substance contains the same number of particles as there are carbon atoms in 12.00 g of carbon-12.

The mass of one mole of an atom is easy to work out. It is simply equal to its relative atomic mass in grams.

$$\text{Amount, } n \text{ (mol)} = \frac{\text{mass (g)}}{\text{molar mass (g mol}^{-1})} = \frac{m}{M}$$

So 24.3 g of magnesium contains 1 mole of atoms, 48.6 g of magnesium contains 2 moles of atoms and so on.

Calculating the amount of substance in moles

Suppose you wanted to know the number of moles in 10.0 g of calcium. You would simply calculate it using amount $= \frac{m}{M}$. The molar mass of calcium is 40.0 g mol^{-1} so amount $= \frac{10.0}{40.0} = 0.25$ mol.

Amount also applies to molecules and ions as well as atoms. For instance, you might want to know the amount of oxygen molecules in 16.0 g of oxygen gas. You would use the same equation but would need to realise the molar mass of oxygen gas, O_2, is 32.0 g mol^{-1}.

So the amount of oxygen molecules $= \frac{16.0}{32.0} = 0.5$ mole.

Avogadro's constant

Experiments have determined the number of atoms in a mole of carbon-12 is 6.02×10^{23}

This value is called the Avogadro constant, N_A.

It can also be applied to other particles. For instance, 1 mole of carbon dioxide will contain 6.02×10^{23} molecules.

Maths skills Calculating number of particles

We can use the Avogadro constant to find the number of particles in a given amount of a substance. Suppose we wanted to know the number of molecules in 8.5 g of ammonia, NH_3.

Firstly we would calculate the amount of ammonia using $\frac{\text{mass}}{\text{molar mass}} \quad \frac{8.5}{17.0} = 0.5$ mol

We can then calculate the number of molecules using this expression.

So the number of molecules $= 0.5 \times 6.02 \times 10^{23}$
$$= 3.01 \times 10^{23} \text{ molecules}$$

Number of particles = amount, n (mol) × Avogadro's constant, N_A (mol^{-1})
$$= n \times N_A$$

Worked example

Calculate the number of moles in
(a) 15 g of $CaCO_3$ **(2 marks)**

Molar mass of $CaCO_3$ = 40.1 + 12.0 + (3 × 16.0) = 100.1 g mol^{-1}.

Amount $= \frac{m}{M} = \frac{15}{100} = 0.15$ mol

(b) 17 g of $(NH_4)_2SO_4$ **(2 marks)**

Molar mass of $(NH_4)_2SO_4$
= [(14.0 + 4.0) × 2] + 32.1 + (4 × 16.0)
= 132.1 g mol^{-1}
Amount $= \frac{m}{M} = \frac{17}{132.1} = 0.13$ mol

Worked example

Calculate the mass of
(a) 1.40 mol of NaOH **(2 marks)**

Molar mass of NaOH = 23.0 + 16.0 + 1.0
= 40 g mol^{-1}

Mass = amount (mol) × M (g mol^{-1})
= 1.40 × 40 = 56 g

(b) 0.050 mol of $KMnO_4$ **(2 marks)**

Molar mass of $KMnO_4$
= 39.1 + 54.9 + (16.0 × 4)
= 158 g mol^{-1}

Mass = amount (mol) × M (g mol^{-1})
= 0.050 × 158 = 7.9 g

Now try this

Calculate the molar mass of a substance for which
(a) 0.28 mol has a mass of 27.4 g **(2 marks)**

(b) 0.62 mol has a mass of 49.6 g. **(2 marks)**

Calculating quantities in reactions

The concept of amount, in moles, can be used to find masses or volumes of substances involved in reactions.

Avogadro's law

Experiments confirmed that at room temperature and pressure, the volume of 1 mole of any gas is $24.0 \, dm^3$. This value is known as the **molar gas volume** and has units of $dm^3 \, mol^{-1}$.

This value can be found on the Data Sheet used in the exam.

Maths skills — Calculating volumes of gases

You need to learn, rearrange and use this key expression as it is useful when looking at reactions involving gases.

$$\text{Amount of gas, } n \text{ (mol)} = \frac{\text{volume of gas, } V \, (dm^3)}{\text{molar gas volume } (dm^3 \, mol^{-1})}$$

This version uses dm^3 as the unit of volume. If the volume of gas is given in cm^3, then the molar gas volume should be converted to $24000 \, cm^3$.

The ideal gas equation

If you know the pressure, temperature and volume of a gas, you can find its amount, in moles, using the ideal gas equation:

$$pV = nRT$$

pressure, in Pa volume in m^3 amount, in mols temperature in K molar gas constant, $8.314 \, J \, mol^{-1} \, K^{-1}$; value given on the Data Sheet in exam

Be careful! Make sure any units used in this type of question are converted to the stated units for the ideal gas equation, e.g. dm^3 to m^3. Don't round up until the last step as this might make your final answer imprecise.

Worked example

1 Calculate the mass, in g, and the volume, in cm^3 of oxygen formed when $3.5 \, g$ of $Mg(NO_3)_2$ is is heated strongly. **(3 marks)**

$$2Mg(NO_3)_2(s) \rightarrow 2MgO(s) + 4NO_2(g) + O_2(g)$$

Amount of limiting reagent $(Mg(NO_3)_2)$

$$= \frac{\text{mass}}{\text{molar mass}} = \frac{3.5}{48.3} = 0.024 \, mol$$

Ratio from equation is 2:1 so amount of

$$O_2 = \frac{0.024}{2} = 0.012 \, mol$$

Mass of O_2 formed = amount × molar mass
$$= 0.012 \times 32 = 0.38 \, g$$

Volume of $O_2(g)$ = amount × molar gas volume
$$= 0.012 \times 24,000 = 288 \, cm^3$$

2 Calculate the relative molecular mass of an oxide of nitrogen, $0.564 \, g$ of which has a volume of $0.426 \, dm^3$ at $100,000 \, Pa$ and $400 \, K$. Use your answer to deduce the formula of the oxide. **(4 marks)**

$$n = pV / RT = \frac{(100,000 \times 4.26 \times 10^{-4})}{(8.314 \times 400)}$$

$$= 0.01281 \, mol$$

Amount in mol, $n = \dfrac{\text{mass}}{\text{relative molecular mass}}$

so relative molecular mass $= \dfrac{\text{mass}}{n}$

$$= \frac{0.564}{0.01281} = 44$$

So oxide of nitrogen must be N_2O

Now try this

Calculate the total volume, in cm^3, of all gases formed when $4.2 \, g$ calcium nitrate, $Ca(NO_3)_2$, is strongly heated.

$$2Ca(NO_3)(s) \rightarrow 2CaO(s) + 4NO_2(g) + O_2(g)$$

(3 marks)

Types of formulae

In simple molecules and giant structures different types of formulae can be used to describe the number or ratio of particles present.

Molecular formulae

Molecular formulae tell you the number and type of each atom in a simple molecule.

White spheres represent hydrogen atoms

Black spheres represent carbon atoms

Red spheres represent oxygen atoms

So, in an ethanol molecule there are 2 carbon atoms, 6 hydrogen atoms and 1 oxygen atom. The molecular formula is C_2H_6O.

Empirical formulae

Empirical formulae show the simplest ratio of atoms of different elements in a compound. These are often used to describe the ratios in giant structures as the actual numbers will depend on the amount of the compound.

In sodium chloride there are many ions present in one small grain. There is always one Na^+ ion to every Cl^- ion so the empirical formula is NaCl.

Cl⁻ ion

Na⁺ ion

Worked example

A compound is found to contain 4.37 g of nitrogen, 5.00 g of oxygen and 0.63 g of hydrogen, by elemental analysis. It has a relative formula mass of 64.0. Calculate its empirical formula and hence deduce its molecular formula.   **(4 marks)**

	Nitrogen	Oxygen	Hydrogen
Amount (mol)	$\frac{4.37}{14.0} = 0.312$	$\frac{5.00}{16.0} = 0.313$	$\frac{0.63}{1.0} = 0.63$
Molar ratio	$\frac{0.312}{0.312} = 1$	$\frac{0.313}{0.312} \approx 1$	$\frac{0.63}{0.312} = 2.02$
Whole number ratio	1	1	2

The empirical formula is NOH_2 with a relative mass of 32.0. The compound's relative formula mass is 64.0, so the molecular formula is double the empirical formula, so $N_2O_2H_4$.

Find the empirical formula by calculating the amount of each element, then divide through by the smallest amount to find the molar ratio. If this is not a whole number ratio scale up as appropriate.

 Maths skills This is the smallest amount, so divide the other amounts by this.

Worked example

Using elemental analysis, compound X is found to contain 40.0% C, 6.7% H and 53.3% O. The molar mass of X is 90. Calculate its empirical formula and hence deduce its molecular formula.   **(4 marks)**

If analysis data is given as percentages a workaround is to assume you have 100 g of the compound and so consider the percentages as masses. You can then work out the amount in moles.

	Carbon	Hydrogen	Oxygen
Amount (mol)	$\frac{400}{12} = 3.33$	$\frac{6.7}{1} = 6.7$	$\frac{53.3}{16} = 3.33$
Molar ratio	$\frac{3.33}{3.33} = 1$	$\frac{6.7}{3.33} \approx 2$	$\frac{3.33}{3.33} = 1$
Whole number ratio	1	2	1

So the empirical formula is CH_2O. Total atomic masses of the atoms in $CH_2O = 12 + 2 + 16 = 30$.

As the molar mass of X is 90, which is 3 times bigger, its molecular formula must be 3 times bigger.

Molecular formula of $X = C_3H_6O_3$.

Now try this

An acid is found to contain 0.50 g of hydrogen, 8.18 g of sulfur and 16.3 g of oxygen. Calculate its empirical formula.   **(3 marks)**

Calculations involving solutions

Many chemical reactions take place in aqueous solution. Calculating chemical amount in such reactions uses volumes and concentrations.

Concentration

Concentration is a measure of the amount of solute dissolved per unit of solvent. In these two beakers the volume of the solvent, water, is the same. Some solute, copper sulfate, is added to each beaker.

Maths skills In this example it's obvious that the solution on the right is more concentrated as its colour is a deeper shade of blue. However, to be useful concentration needs to be quantified using this equation:

$$\text{concentration, } c \text{ (mol dm}^{-3}) = \frac{\text{amount, } n \text{ (mol)}}{\text{volume, } V \text{ (dm}^3)}$$

A small amount of copper sulfate is added – this solution is dilute.

A larger amount of copper sulfate is added – this solution is more concentrated.

Worked example

1 Calculate the concentration of the solution formed when 0.20 mol of NaCl(s) is added to 400 cm³ of water.   **(1 mark)**

Concentration, $c = \dfrac{n}{V}$ 1000 cm³ = 1 dm³,

so 400 cm³ = $\dfrac{400}{1000}$ = 0.4 dm³

So $c = \dfrac{0.20}{0.4}$ = 0.50 mol dm⁻³

As concentration has units of mol dm⁻³, the volume used in the calculation must be in dm³.

2 Calculate the concentration of the solution formed when 16.0 g of NaOH is dissolved in 200 cm³ of water.   **(2 marks)**

Amount of NaOH = $\dfrac{\text{mass}}{\text{molar mass}}$

$= \dfrac{16.0}{40.0}$ = 0.40 mol

$V = \dfrac{200}{1000}$ = 0.2 dm³   Remember that V should be in dm³.

concentration $c = \dfrac{n}{V}$

$= \dfrac{0.40}{0.2}$ = 2.0 mol dm⁻³

3 Calculate the mass of Na₂CO₃ needed to make 500 cm³ of a 0.016 mol dm⁻³ solution and hence deduce its concentration in g dm⁻³.   **(3 marks)**

Amount of Na₂CO₃ needed $n = c \times V$

So $n = 0.016 \times \left(\dfrac{500}{1000}\right)$ = 0.0080 mol

Mass of Na₂CO₃ = amount × molar mass

$= 0.008 \times 106.0 = 0.85$ g

1 dm³ is equal to 1000 cm³, so if 0.85 g are required in 500 cm³, 1.70 g would be needed in 1000 cm³, i.e. 1.70 g dm⁻³.

Now try this

Calculate the volume of a 0.20 mol dm⁻³ solution of nitric acid needed to form 50 cm³ of CO_2(g) in the reaction with excess K_2CO_3.

$$K_2CO_3(s) + 2HNO_3(aq) \rightarrow 2KNO_3(aq) + H_2O(l) + CO_2(g)$$ **(3 marks)**

Formulae of hydrated salts

Hydrated salts have water molecules attached to the ions in their crystal structure, called water of crystallisation. Simple experiments can be carried out to find the ratio of water to salt and hence the formula, including the water of crystallisation.

 Heating hydrated copper sulfate

As a sample of copper sulfate is heated, the water molecules will break away from the crystal structure, forming water vapour. If heated for sufficient time all the water will be removed, leaving anhydrous copper sulfate. By carefully recording the mass of the hydrated salt before heating and the anhydrous salt after heating, the results can be processed to find the ratio of salt to water.

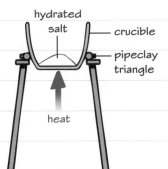

Hydrated copper sulfate
→ anhydrous copper sulfate + water
$CuSO_4 \cdot xH_2O(s) \rightarrow CuSO_4(s) + xH_2O(g)$

Using the results to determine the formula of hydrated copper sulfate

	Mass / g
Crucible	10.48
Crucible and hydrated copper sulfate	12.63
Crucible and anhydrous copper sulfate	11.85

1 Mass of anhydrous copper sulfate formed = 11.85 − 10.48 = 1.37 g

Mass of water removed = 12.63 − 11.85 = 0.78 g

2 Amount of anhydrous copper sulfate
$= \dfrac{mass}{M_r} = \dfrac{1.37}{159.6} = 0.0086$ mol

Amount of water $= \dfrac{mass}{M_r} = \dfrac{0.78}{18.0}$
$= 0.043$ mol

3 Ratio of $CuSO_4 : H_2O = 0.0086 : 0.043$

Divide through by smallest number to find whole number ratio = 1 : 5

4 Formula of hydrated copper sulfate is $CuSO_4 \cdot 5H_2O$

Worked example

5.46 g of hydrated sodium carbonate, $Na_2CO_3 \cdot xH_2O$, was heated gently to remove the water of crystallisation. After cooling 2.02 g of anhydrous sodium carbonate remained. Use these results to find the formula of the hydrated salt. **(4 marks)**

Firstly, calculate the mass of the water removed:

5.46 − 2.02 = 3.44 g

Then calculate the amount of anhydrous salt and the amount of water:

Amount of anhydrous salt $= \dfrac{mass}{M_r} = \dfrac{2.02}{106.0} = 0.019$ mol

Amount of water $= \dfrac{mass}{M_r} = \dfrac{3.44}{18.0} = 0.19$ mol

Finally, deduce the whole number ratio and hence formula:

$0.019 : 0.19 \equiv 1 : 10$, so formula is $Na_2CO_3 \cdot 10H_2O$

Now try this

4.74 g of a hydrated salt with the formula $XCl_2.2H_2O$ was heated, forming 3.58 g of the anhydrous salt. Use the data to find the relative formula mass of the salt and hence identify X. **(4 marks)**

Remember if you cannot spot the whole number ratio by just looking at the values, divide through by the smallest value.

Percentage yield and atom economy

Percentage yield and atom economy are different ways of looking at the efficiency of a chemical reaction in terms of the amount of product formed.

Percentage yield

Calculations to work out the amount of product formed in a reaction assume that *all* the reactants are converted to products. This would give a yield of 100%. In practice this is very difficult to achieve because:

- the reaction may be reversible
- the reactants used may contain impurities
- by-products may be formed from unexpected reactions
- reactants or products may be left in apparatus during the experiment
- product may be lost during separation and purification, such as being soaked onto filter paper.

Atom economy

Atom economy considers the efficiency of a reaction by looking at the proportion of the products formed that are actually useful.

For instance we can use the reaction between ethene and steam to produce the desired product, ethanol:

$C_2H_4(g) + H_2O(g) \rightarrow C_2H_5OH(l)$

Hence this reaction has an atom economy of 100% as all the reactants are converted to the desired product.

However when calcium oxide is produced from calcium carbonate, the carbon dioxide formed is a waste product:

$CaCO_3(s) \rightarrow CaO(s) + CO_2(g)$

Hence this reaction has an atom economy of less than 100%.

Reactions that have a high atom economy are increasingly important as

- they make better use of finite resources
- they reduce the need to process and safely dispose of waste products.

Worked example

(a) A student used 0.200 mol of CuO with excess H_2SO_4 forming 31.4 g of $CuSO_4 \cdot 5H_2O$. Calculate the percentage yield.   **(3 marks)**

$CuO + H_2SO_4 + 5H_2O \rightarrow CuSO_4 \cdot 5H_2O + H_2O$

This means 0.200 mol of CuO should produce a theoretical yield 0.200 mol of anhydrous $CuSO_4 \cdot 5H_2O$ (1:1 ratio in equation).

Actual amount produced $= \dfrac{mass}{M_r}$

$= \dfrac{31.4}{249.6}$

$= 0.126$ mol

So % yield $= \dfrac{0.126}{0.200} = 63.0\%$

(b) The water produced in the reaction on the left is not a useful product. Calculate the atom economy of the reaction.   **(2 marks)**

Atom economy

$= \dfrac{\text{molecular mass of desired product}}{\text{sum of molecular masses of all products}}$

$= \dfrac{249.6}{(249.6 + 18.0)} \times 100$

$= 93.3\%$

Definition of percentage yield

Percentage yield

$= \dfrac{\text{actual amount (mol) of product}}{\text{theoretical amount (mol) of product}}$

Definition of atom economy

Atom economy

$= \dfrac{\text{molecular mass of desired product}}{\text{sum of molecular masses of all products}}$

Remember that atom economy and percentage yield are not the same thing! A reaction with a high atom economy is not necessarily the most sustainable way to produce a product as that reaction may have a very low percentage yield. The challenge for chemists is to research processes that combine high atom economy with a good yield.

Now try this

1 10.0 g of C_2H_4 reacted with excess H_2O forming 13.7 g of C_2H_5OH. Calculate the percentage yield.   **(3 marks)**

2 An alternative way to make C_2H_5OH is fermentation.
$C_6H_{12}O_6 \rightarrow 2C_2H_5OH + 2CO_2$
Calculate the atom economy of fermentation.   **(2 marks)**

Neutralisation reactions

Acids can be neutralised by a range of compounds, called bases, to form salts.

Common acids

Common acids you need to know are:

- hydrochloric acid, HCl(aq)
- sulfuric acid, H_2SO_4(aq)
- nitric acid, HNO_3(aq).

They all dissociate to release hydrogen ions, H^+(aq), when dissolved in water. As an H^+ ion is simply a proton, acids can be defined as proton donors. The H^+ ion is responsible for the reactions of acids.

Hydrochloric acid, HCl(aq), is an example of a **strong acid**.

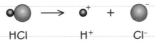

HCl → H⁺ + Cl⁻

When it dissolves **all** the HCl molecules dissociate to release hydrogen ions.

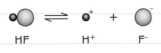

HF ⇌ H⁺ + F⁻

Hydrofluoric acid, HF(aq), is an example of a weak acid.

When it dissolves **only a small proportion** of the HF molecules dissociate to release hydrogen ions.

Common bases

Common bases you need to know are:

- potassium hydroxide, KOH
- sodium hydroxide, NaOH
- ammonia, NH_3.

Other examples include metal oxides and metal carbonates.

Bases that dissolve in water are also known as **alkalis**. They dissociate to release hydroxide ions, OH^-(aq), when dissolved in water.

Bases **neutralise** acids by accepting the protons donated by the acid. For example, the hydroxide ion from sodium hydroxide will accept a H^+ ion to form water.

$$H^+(aq) + OH^-(aq) \rightarrow H_2O(l)$$

This means bases can be defined as **proton acceptors**.

Sodium hydroxide is a **strong base,** as when it dissolves **all** of its hydroxide ions are available to accept protons.

Ammonia is a **weak base** as **only a small proportion** of the NH_3 molecules will accept protons.

Forming salts by neutralising acids with bases

When hydrogen ions in an acid are replaced by metal ions from a base, a salt is formed. Examples of bases include carbonates, metal oxides and alkalis. Such reactions are called **neutralisations**.

> This is really two definitions in one question – you should include what defines an acid and what makes it strong.

Acid reactions

Example of an acid reaction with a metal oxide:

magnesium oxide + nitric acid → magnesium nitrate + water

$$MgO(s) + 2HNO_3(aq) \rightarrow Mg(NO_3)_2(aq) + H_2O(l)$$

Example of an acid reaction with an alkali:

potassium hydroxide + sulfuric acid → potassium sulfate + water

$$2KOH(s) + H_2SO_4(aq) \rightarrow K_2SO_4(aq) + H_2O(l)$$

Example of an acid reaction with a carbonate:

calcium carbonate + hydrochloric acid
→ calcium chloride + water + carbon dioxide

$$CaCO_3(s) + HCl(aq) \rightarrow CaCl_2(aq) + H_2O(l) + CO_2(g)$$

Define the term strong acid.   **(2 marks)**

A strong acid is a substance that fully dissociates to donate hydrogen ions dissolved in water.

> Don't confuse the strength of an acid with its concentration. Concentration is the amount of acid dissolved to make 1 dm³ of weak solution.

1 Describe how a hydrogen ion bonds to a hydroxide ion, stating the name of the type of bond formed.   **(2 marks)**

2 Write a balanced equation for each of these neutralisation reactions:
   (a) sodium carbonate with hydrochloric acid
   (b) ammonia with sulfuric acid
   (c) copper oxide with nitric acid.   **(6 marks)**

Acid–base titrations

Practical skills An acid–base titration is an experiment in which an acid solution is carefully added to a basic solution, or vice versa, until the reactants just neutralise each other. An indicator is often used to denote this point. If the concentration of one solution is known, the volumes used in the experiment can be processed to find an unknown concentration.

Setting up a titration

A **standard solution** is one whose concentration is known accurately. They are often used in experiments called **titrations** to find the concentration of another solution.

Volume is read from bottom of the meniscus and to 2 d.p. The second d.p. is estimated to either a 0 or a 5.

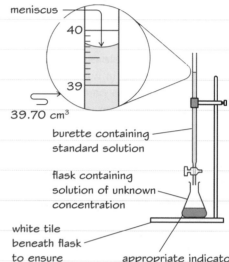

meniscus

39.70 cm³

burette containing standard solution

flask containing solution of unknown concentration

white tile beneath flask to ensure colour change shows up

appropriate indicator used to highlight end point of titration

- Burette set up containing standard solution.
- Initial volume noted.
- Accurate known volume of other solution added to flask, using a volumetric pipette.
- Small amount of indicator added to flask.
- Standard solution added slowly to flask.
- Flask swirled gently to ensure mixing.
- Sides of flask rinsed with distilled water if needed.
- Stop adding when indicator **just** changes permanently.
- Final volume noted.

Common indicators used in acid-base titrations

Indicator	Colour in acid	Colour in base	End point colour
methyl orange	red	yellow	orange
bromothymol blue	yellow	blue	green
phenolphthalein	colourless	pink	pale pink

The end point colour for phenolphthalein assumes base is being added to acid. If the experiment is carried out with acid being added to base the colour at the end point will be colourless.

Different types of errors can affect experiments such as titrations, for example:

- Systematic error – a constant error due to a piece of equipment, e.g. a balance used for weighing out solid samples being incorrectly calibrated.
- Random error – an error that may not always occur in an experiment, e.g. sometimes leaving an air bubble in the burette.
- Margin of error of equipment – limits of precision of equipment used to take measurements, e.g. precision of a pipette to ±0.06 cm³.

Worked example

(a) A 25.0 cm³ volumetric pipette was used in a titration to measure out a sample of $NaOH(aq)$. The margin of error of the pipette was ±0.06 cm³. Calculate the percentage error. **(1 mark)**

Percentage error of equipment

$$= \left(\frac{\text{maximum margin of error}}{\text{measured value}} \right) \times 100$$

$$= \frac{0.06}{25.0} \times 100 = 0.24\%$$

(b) A student found by experiment that the solubility of a salt was 1.45 g dm⁻³. The true value from a reliable data book was 1.68 g dm⁻³. Calculate the percentage error in the student's value. **(1 mark)**

Percentage error of value

$$= \left[\frac{|\text{true value} - \text{experimental value}|}{\text{true value}} \right] \times 100$$

$$= \left[\frac{|1.68 - 1.45|}{1.68} \right] \times 100 = 13.7\%$$

Now try this

Explain why rinsing the sides of the flask during a titration does **not** affect the final volume required (the titre). **(2 marks)**

Calculations based on titration data

📇 **Maths skills** Analysing titration data allows you to determine an unknown property of the solution you are testing. This is normally concentration, but can include properties such as molar mass.

Analysing titration data

In a titration, a student wanted to find out the concentration of some NaOH(aq). A $25.0\,cm^3$ sample of this solution was titrated with a standard solution of HCl(aq) of concentration $0.105\,mol\,dm^{-3}$. The reaction equation is:

$NaOH(aq) + HCl(aq) \rightarrow NaCl(aq) + H_2O(l)$

Before starting any calculations, summarise all the data in the problem.

- average volume of HCl(aq) $= \dfrac{19.15 + 19.05}{2}$

$\qquad = 19.10\,cm^3$

- concentration of HCl(aq) $= 0.105\,mol\,dm^{-3}$
- volume of NaOH(aq) used $= 25.0\,cm^3$
- unknown concentration of NaOH(aq).

All the readings from the burette are given to two decimal places but the second decimal place is always a '0' or '5'. This is because burettes are accurate to $\pm 0.05\,cm^3$.

Experiment	Volume before (cm^3)	Volume after (cm^3)	Titre (cm^3)
1	0.00	19.15	19.15
2	19.10	38.55	19.45
3	0.00	19.05	19.05

These two titres are within $0.10\,cm^3$ of each other so are **concordant**. They will be used to calculate the average titre.

Average titre $= \dfrac{(19.15 + 19.05)}{2} = 19.10\,cm^3$

Calculating the concentration of the NaOH(aq):

1 Calculate the amount of the standard solution that reacted in the titration.

2 Deduce the amount of unknown solution that reacted in the titration.

3 Calculate the concentration of the solution that reacted in the titration using the amount and volume used in the titration.

This is HCl(aq) and we can find the amount by:

moles = volume (in dm^3) × concentration

$= \dfrac{19.10}{1000} \times 0.105 = 2.0055 \times 10^{-3}\,mol$

This is NaOH(aq) and we find the amount by using the ratio NaOH : HCl from the balanced equation, which is 1:1.

moles of NaOH = moles of HCl

$= 2.0055 \times 10^{-3}\,mol$

concentration $= \dfrac{\text{amount (mol)}}{\text{volume (dm}^3)} = \dfrac{2.0055 \times 10^{-3}}{\frac{25}{1000}} = 0.080\,mol\,dm^{-3}$

Worked example

A student made $100\,cm^3$ of solution of an acid H_2X by dissolving $1.96\,g$ of H_2X in distilled water. $20.0\,cm^3$ samples of this solution were titrated with a solution of NaOH(aq) of concentration $0.400\,mol\,dm^{-3}$. The average titre was $20.1\,cm^3$. Calculate the M_r of the acid. **(4 marks)**

$\dfrac{20.1}{1000} \times 0.400$, amount of NaOH $= 8.04 \times 10^{-3}$

So the amount of H_2X in $20.0\,cm^3 = 4.02 \times 10^{-3}$

Originally solid was dissolved to make $100\,cm^3$ of solution, so total amount of H_2X dissolved $= 4.02 \times 10^{-3} \times 5 = \mathbf{0.0201\,mol}$

$M_r = \dfrac{\text{mass}}{\text{amount (in mol)}} = \dfrac{1.96}{0.0201} = \mathbf{97.5\,g\,mol^{-1}}$

Finally the M_r of H_2X can now be found as we know both the amount in moles and the mass in grams.

Firstly calculate the amount of NaOH that reacted.

$H_2X + 2NaOH \rightarrow Na_2X + 2H_2O$

Find the ratio of $H_2X : NaOH$ and deduce the amount of H_2X. As each molecule of H_2X will release 2 hydrogen ions the ratio is $2NaOH : 1\,H_2X$

Now try this

Find the concentration of a solution of nitric acid, HNO$_3$(aq), $20.0\,cm^3$ of which required $17.8\,cm^3$ of $0.250\,mol\,dm^{-3}$ KOH(aq) to neutralise it in a titration.   **(3 marks)**

Oxidation numbers

Chemists use oxidation numbers to help determine which particles are oxidised and which are reduced, and to help construct formulae and balance equations.

Rules for assigning oxidation numbers

Some elements have oxidation numbers that rarely vary in compounds. These values can be used to determine the oxidation state of those elements whose oxidation numbers do vary.

✓ Uncombined elements have an oxidation number of 0.

✓ The sum of all oxidation numbers in a compound is 0.

✓ The sum of all oxidation numbers in an ion equals the charge on that ion.

✓ Group 1, 2 and 3 elements in compounds have oxidation number +1, +2 and +3 respectively.

✓ Fluorine has the oxidation number −1 in all its compounds.

✓ Hydrogen has the oxidation number +1 in its compounds, except metal hydrides (when it is −1).

✓ Oxygen has the oxidation number −2 in most of its compounds. The exceptions are compounds with fluorine, peroxides and superoxides, e.g. hydrogen peroxide, H_2O_2, in which oxygen has the oxidation number −1.

✓ Chlorine has the oxidation number −1 in most of its compounds. The exceptions are compounds with fluorine and oxygen, when it can vary. For instance in NaClO, chlorine has the oxidation number +1.

Finding oxidation numbers

In this example you can use the rules to find the oxidation number of sulfur in Na_2SO_4.

The oxidation number of sulfur varies in different compounds so has to be deduced using the known values for the other elements.

Each Na has an oxidation number of +1, so +2 in total for Na.

The oxidation number of S is +6 in this compound

$$Na_2SO_4$$

Each O has an oxidation number of −2, so −8 in total for O.

Worked example

Deduce the oxidation number of each element in the compound KIO_3. **(2 marks)**

+1 for K and −2 for each O. However there are 3 oxygen atoms so that gives a total of (3 × −2) = −6. So the total due to K and O is −5. Hence the oxidation number of I in KIO_3 is +5.

Using oxidation numbers in redox reactions

By deducing the oxidation numbers of the species involved, you can summarise the changes in a redox reaction. For example,

Zinc has been oxidised from 0 to +2.

Hydrogen in HCl has been reduced from +1 to 0.

$$Zn + 2HCl \rightarrow ZnCl_2 + H_2$$
$$0 {+1}{-1} {+2}{-2} 0$$

The changes in oxidation numbers balance for both the reduction and oxidation parts of the reaction. Each Zn atom is oxidised from 0 to +2. To balance this, **two** hydrogen ions are reduced by −1 each to give an overall reduction of −2.

Now try this

1 Deduce the oxidation numbers of each element in $KMnO_4$. **(2 marks)**

2 Use oxidation numbers to confirm what has been oxidised, and what has been reduced, in this redox reaction:

$$2HBr + H_2SO_4 \rightarrow Br_2 + SO_2 + 2H_2O$$
(2 marks)

Examples of redox reactions

Oxidation numbers are used to interpret what is happening in reactions you are not familiar with.

Oxidation numbers in names

Some elements, such as transition metals, can have different oxidation states in compounds or ions (see page 14). The oxidation number, written as a Roman numeral, is given in the name of the compound.

The oxidation number of Fe varies in different compounds. Here it must be +3 to balance the -3 of the Cl.

$FeCl_3$

Each Cl has an oxidation number of -1 so -3 in total for Cl.

This means the compound must be called iron(III) chloride.

Using oxidation numbers to interpret an unfamiliar redox reaction

Working out changes in oxidation number can help to deduce which species are oxidised and which are reduced. They can also help to balance the equation for the reaction. In this example oxidation numbers are used to work out whether the nitrogen or carbon has been oxidised in the reaction:

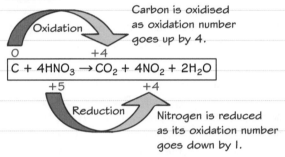

Carbon is oxidised as oxidation number goes up by 4.

Oxidation

$$0 \qquad\qquad +4$$
$$C + 4HNO_3 \rightarrow CO_2 + 4NO_2 + 2H_2O$$
$$+5 \qquad\qquad +4$$

Reduction

Nitrogen is reduced as its oxidation number goes down by 1.

Four NO_2 molecules need to react to balance the equation. This ensures the total change in oxidation number is the same for both the reduction and the oxidation parts of the reaction.

Identify the changes in oxidation number in this reaction. What is unusual about these changes?

$$Cl_2 + 2OH^- \rightarrow ClO^- + Cl^- + H_2O \qquad \textbf{(2 marks)}$$

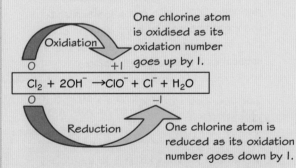

One chlorine atom is oxidised as its oxidation number goes up by 1.

Oxidiation

$$0 \qquad\qquad +1$$
$$Cl_2 + 2OH^- \rightarrow ClO^- + Cl^- + H_2O$$
$$0 \qquad\qquad -1$$

Reduction

One chlorine atom is reduced as its oxidation number goes down by 1.

This is an unusual reaction as atoms of one element in the same species are oxidised and reduced at the same time. This is called a **disproportionation** reaction.

Use oxidation numbers to name these two compounds.
(a) Na_2SO_4
The oxidation number of S in Na_2SO_4 is +6, so the name is sodium sulfate(VI).
(b) Na_2SO_3.   **(2 marks)**

The oxidation number of S in Na_2SO_3 is +4, so the name is sodium sulfate(IV).

Deduce any changes of oxidation number in this reaction and use them to help balance the equation.

$$MnO_4^- + H^+ + Fe^{2+} \rightarrow Mn^{2+} + Fe^{3+} + H_2O$$
     (2 marks)

Remember that the changes in oxidation numbers have to balance.

Exam skills 1

This exam-style question uses knowledge and skills you have already revised. Have a look at pages 8, 12 and 13 for a reminder about **concentration** and **titrations**, including calculations.

Worked example

A solution of a metal hydroxide, MOH, is prepared using a volumetric flask.

The procedure followed is

1. Measure the solid in a weighing boat using a balance accurate to 0.01 g.
2. Add the solid to a volumetric flask using a funnel.
3. Add water to the volumetric flask until the volume of the solution is exactly 100 cm³.

(a) Suggest two changes that will improve the accuracy of this procedure. Explain each of your suggestions.   **(4 marks)**

One way to improve this part of the procedure is to reweigh the weighing boat after the solid is added to the flask and then find the difference in mass. This tells you how much solid has been added to the flask. This means that any solid left in the weighing boat is taken into account.

Also the procedure does not refer to stirring or shaking as the solution is made. This is important to ensure the solid dissolves completely to form a uniform solution.

(b) A student weighs out 0.14 g of MOH to prepare such a solution. 25.0 cm³ of this solution is titrated using hydrochloric acid, HCl, of concentration 0.050 mol dm⁻³. 17.20 cm³ of the hydrochloric acid is needed to neutralise the MOH solution. Use this information to deduce the identity of the metal, M.   **(5 marks)**

$MOH + HCl \rightarrow MCl + H_2O$
Amount of HCl = (17.20 ÷ 1000) × 0.050
          = 8.6 × 10⁻⁴ mol
So, we can deduce that the amount of MOH (in 25.0 cm³) is 8.6 × 10⁻⁴ mol as the ratio in the equation is 1:1.
Hence the amount of MOH in 100 cm³
= 8.6 × 10⁻⁴ × 4 = 3.44 × 10⁻³ mol
Relative formula mass of MOH = mass ÷ amount
= 0.14 ÷ (3.44 × 10⁻³) = 40.7
So, subtracting 17.0 (for relative masses of O and H atoms) we are left with 23.7
Hence M must be Na (relative atomic mass = 23.0)   **(5 marks)**

(c) Explain whether or not the student's results are reliable.   **(2 marks)**

The results are not reliable as there is no evidence that the titration was carried out more than once.

Practical skills   You can expect to see questions that test your knowledge of key practical techniques such as titrations. Often they will expect you to consider how to improve a procedure.

'Suggest' means that you may not have been taught this idea directly, but you should apply your knowledge to an unknown situation. In this case you will have to read the procedure and consider whether any key information has been omitted or described incorrectly.

Another way to improve the transfer of the solid to the volumetric flask would be to add the solid to a small beaker, dissolve in distilled water and then transfer to the flask, then rinsing, making sure all rinsings go into the volumetric flask.

Other improvements could include more detailed references to the way equipment is used, e.g. reading volumetric flask at eye level.

Try to practise as many different types of questions based on titrations as possible. Start with examples similar to the experiments you have done in class then build up to tackling some of the more unusual examples on past papers.

Maths skills   Carefully show your working so that if you make a slip the examiner can follow what you have done and award part-marks. This is especially important in calculations such as this example, without a clear structure to guide you through, so try to label each step to help them see what you are doing.

Practical skills   Reliability in titrations is determined by obtaining concordant results, which means two titre values within ±0.10 cm³.

Electron shells and orbitals

The 'shell' model describes the shapes in space where the electrons in an atom are concentrated.

Main energy levels or shells

Electrons orbit the nucleus in main energy levels, or shells. The further the shell is from the nucleus, the higher its energy. Each shell is given a number, called a **principal quantum number**, and can hold the maximum number of electrons.

n	Shell	Number of electrons
1	1st shell	2
2	2nd shell	8
3	3rd shell	18
4	4th shell	32

Atomic orbitals

Each shell consists of atomic orbitals. These are regions in space where electrons may be found. Each orbital contains a maximum of 2 electrons. The first four types of orbital are called s-, p-, d- and f-orbitals.

The diagram shows that the first shell has one type of orbital, the second shell has two types of orbitals and so on. Orbitals of the same type within a shell are known as **sub-shells**.

The number of orbitals in a sub-shell is summarised in the table.

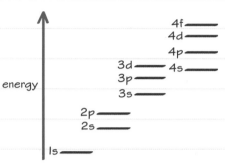

Sub-shell	Number of orbitals
s	1
p	3
d	5
f	7

Shapes of orbitals

An s-orbital is spherical in shape and holds up to 2 electrons.

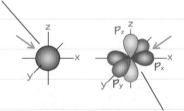

Each p-orbital is dumb-bell shaped and holds up to 2 electrons. Each p sub-shell has 3 p-orbitals, p_x, p_y and p_z

Box diagrams

It's helpful to represent each orbital as a box and the electrons as arrows. The direction of the arrow shows the **spin** of the electron. Two electrons in the same orbital spin in opposite directions to offset their tendency to repel each other.

 s sub-shell, with 1 box to show the single orbital.

 p sub-shell, with 3 boxes to show each orbital

Worked example

What is the maximum number of electrons that a p-orbital can hold? **(1 mark)**

2 electrons

Use the 'electrons in boxes' model to draw a full d sub-shell, which can hold up to 10 electrons. **(2 marks)**

Do not confuse 'p-orbital' with 'p sub-shell'. A p sub-shell consists of three p-orbitals, so can hold a maximum of six electrons, but each individual p-orbital can only hold a maximum of two electrons.

A maximum of 10 electrons must mean five d-orbitals holding two electrons each, so 5 boxes are shown. The arrows in each box must be in opposite directions to denote the opposite spins.

Now try this

Try squaring the principal quantum number.

Deduce the relationship between the principal quantum number of an energy level and the maximum number of electrons it can hold. **(1 mark)**

Electron configurations – filling the orbitals

The model of electron orbitals, sub-shells and main shells can be used to write electron configurations and 'electrons in box' representations of configurations.

Writing electron configurations

Electron configurations are the arrangement of the electrons in atoms or ions. An advanced model recognises that the main levels are split into sub-levels so our electron configurations should reflect that.

The standard way of representing electron configurations is:

Element	Electron configuration
B	$1s^2 2s^2 2p^1$
C	$1s^2 2s^2 2p^2$
N	$1s^2 2s^2 2p^3$
O	$1s^2 2s^2 2p^4$

Electrons in boxes for the elements B, C, N and O

These diagrams show electrons in boxes. Each box represents an orbital, so the p sub-shell has 3 boxes. The arrows represent electrons in the orbitals. They are helpful as they show the number of unpaired electrons which is useful when looking at bonding.

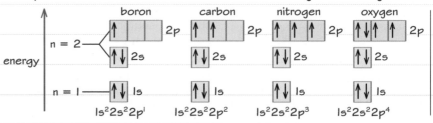

The large numbers represent the energy levels, the letter the sub-level and the superscript the number of electrons in that sub-level.

Rules for arranging electrons in electron configurations

- Start at the lowest energy level and add electrons one at a time to build up the configuration.
- Fill each sub-shell before starting on the next.
- When using electrons in boxes put an electron in each orbital singly within a sub-shell, before starting to pair electrons.
- Paired electrons have opposite spins, so when using 'electrons in boxes' they are shown as arrows pointing in opposite directions.

Watch out for 3d and 4s!

You need to be able to draw electron configurations for the first 36 elements and as you reach the higher energy levels the sub-shells get closer together, or converge.

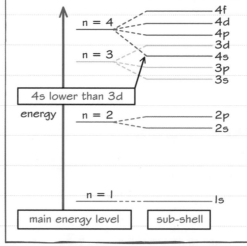

After the 3p sub-shell is filled, the 4s sub-shell is filled as it is lower in energy than the 3d sub-shell. However once occupied with electrons, 4s moves to a slightly higher level. So, when forming ions electrons are removed from the 4s sub-shell before the 3d sub-shell.

Now try this

Draw 'electron in boxes' representations of the electrons in the following:

(a) Cl atom

(b) Ni²⁺ ion

(c) Ge atom. **(3 marks)**

Don't forget to change the number of electrons to take into account that this is an ion.

Worked example

Write the electron configuration of the following:

(a) Na atom 11 electrons so $1s^2 2s^2 2p^6 3s^1$

(b) O^{2-} ion 10 electrons so $1s^2 2s^2 2p^6$

(c) Ti atom 22 electrons so $1s^2 2s^2 2p^6 3s^2 3p^6 3d^2 4s^2$

(d) V^{3+} ion 20 electrons so $1s^2 2s^2 2p^6 3s^2 3p^6 3d^2$

(4 marks)

Ionic bonding

Ionic bonds are electrostatic attractions between positive and negative ions. Ions formed in an ionic compound can be represented using 'dot-and-cross' diagrams.

Ionic compounds and giant ionic structures

Ionic compounds have giant structures.

In a giant ionic structure the ions are arranged in a regular, three-dimensional pattern called a **lattice**. The electrostatic forces between the ions act in all directions and keep the structure together. The large number of these strong electrostatic attractions means that ionic compounds have high melting points.

Sodium chloride lattice

In the sodium chloride lattice each sodium ion is surrounded by six chloride ions and each chloride ion is surrounded by six sodium ions. This repeating pattern continues for a vast number of ions.

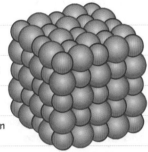

Cl⁻ ion

Na⁺ ion

Positive ions

- Positive ions generally form by metal atoms losing electrons.

- In Groups 1, 2 and 3 the positive charge of the ions formed is equal to the group number of the element.

- Transition metals can often form more than one ion with different charges (e.g. Fe^{2+}, Fe^{3+}).

- Positive ions can be represented in a dot-and-cross diagram, for example a Na^+ ion.

Outer shell now empty as sodium atom has lost one electron to become an ion.

Must show charge on ion.

Square brackets to show charge is spread over whole ion.

Negative ions

- Negative ions generally form by non-metal atoms gaining electrons from metal ions.

- In Groups 7, 6 and 5 the magnitude of the negative charge of the ions formed is equal to 8 minus the group number of the element.

- It's helpful to learn the charges on polyatomic ions such as CO_3^{2-}, SO_4^{2-}, NO_3^- and OH^-.

- Negative ions can be represented in a dot-and-cross, diagram, for example a F^- ion.

One of the electrons is shown as a cross to show that it has been gained by a fluorine atom.

Draw a dot-and-cross diagram to show the bonding in calcium chloride. Show the outer electrons only.
(2 marks)

Notice the command to show outer electrons only. This is actually to make life easier for you!

However, it can confuse some students who don't know whether to show the empty outer shell in the calcium ion or the last-but-one shell of the atom, with eight electrons, which could be considered the new outer shell of the ion. Both are correct!

Also in this example the non-metal electrons are shown as crosses. This is fine – there is no hard and fast rule.

Notice that both chloride ions have been drawn, though it's acceptable to draw one but indicate that two are needed, e.g. by using 'x2'. Each chloride ion has a dot to show the electron gained from the calcium atom.

Now try this

1 Draw a dot-and-cross diagram to show the bonding in sodium nitride. Show the outer electrons only. **(2 marks)**

2 Sodium and magnesium ions are said to be *isoelectronic*. Draw dot-and-cross diagrams of these ions and use them to suggest what isoelectronic means. **(3 marks)**

Covalent bonds

A covalent bond is an electrostatic attraction between a shared pair of electrons and the nuclei of the bonded atoms.

How do atoms form covalent bonds?

A covalent bond forms when atoms share a pair of electrons. Generally each atom in the bond contributes one electron to the pair, but a covalent bond consisting of an electron pair derived from one of the atoms is called a **dative** covalent bond or a **coordinate** bond.

'Dot-and-cross' diagrams

These are a way of showing the bonding between atoms, for instance in a chlorine molecule.

Dots represent electrons from one atom, crosses from the other.

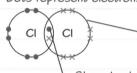

Circles represent the shells. Only the outer electrons need to be shown.

Shared pair in overlapping shells between the atoms represents the covalent bond.

Worked example

1 Draw a dot-and-cross diagram of NH_3. (1 mark)

Each hydrogen atom has I unpaired electron (Is¹) in its outer shell, so can form a single bond.

The pair of non-bonded electrons on the nitrogen atom is called a **lone pair**.

The nitrogen atom has 3 unpaired electrons (2p³) in its outer shell, so can form 3 single bonds

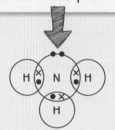

2 Draw a dot-and-cross diagram of BCl_3. (1 mark)

The boron atom has 3 unpaired electrons in its outer shell, so can form 3 single bonds.

Notice that the boron atom only has 6 electrons in its outer shell, as all its outer shell electrons are paired.

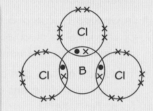

Each chlorine atom has I unpaired electron in its outer shell, so can form I single bond.

3 Draw a dot-and-cross diagram of CO_2. (1 mark)

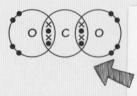

Each oxygen atom shares 2 pairs of electrons with the carbon atom. Each oxygen forms a **double** bond.

4 Draw a dot-and-cross diagram of NH_4^+. (1 mark)

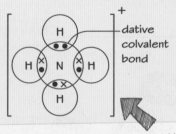

dative colvalent bond

The NH_4^+ ion forms from NH_3 and H^+.

The H^+ ion has no electrons, but the nitrogen atom has a lone pair not involved in bonding. It uses this to bond to the H+, forming a **dative covalent bond**.

Now try this

1 Draw a dot-and-cross diagram of the compound $BF_3 \cdot NH_3$. **(2 marks)**
2 Draw a dot-and-cross diagram of the OH^- ion. **(1 mark)**

Shapes of molecules

You can predict the shapes of simple molecules by looking at the number and type of electron pairs around a central atom and considering how those pairs will repel each other – electron pair repulsion theory.

Electron pair repulsion theory

This theory can be used to predict the shape of simple molecules, and of angles within them.
To find the shape:

✓ Draw a dot-and-cross diagram to find the number and type of electron pairs around the central atom.

✓ Bonding pairs of electrons will repel each other equally and give the shape and angle that maximises the distance between them.

✓ If there are only two or three electron pairs the shapes will be *planar* (flat).

✓ More than three electron pairs will form a 3D shape.

✓ Electrons in double bonds exist as one area of electron density so repel in the same way as the electrons in a single bond.

The table summarises the shapes of molecules with up to six electron pairs around the central atom.

Number of bonded electron pairs around central atom	1	2	3	4	5	6
Name of shape	linear	linear	trigonal planar	tetrahedral	trigonal bipyramid	octahedral
Example molecule	H_2	CO_2	BF_3	CH_4	PCl_5	SF_6
Dot-and-cross diagram	 H×H	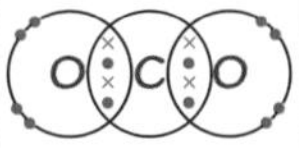 O⦂C⦂O	 F, B, F, F	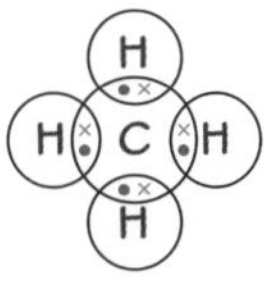 H C H (×4)	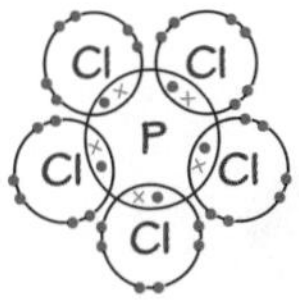 Cl, Cl, P, Cl, Cl, Cl	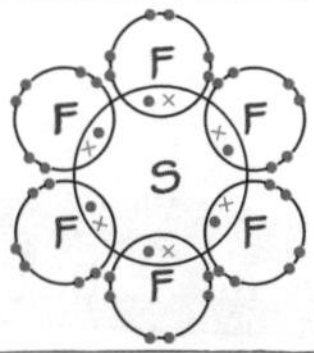 F, F, F, S, F, F, F
Shape and bond angles	 H—H	 O=C=O 180°	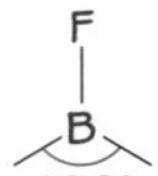 F|B<F F 120°	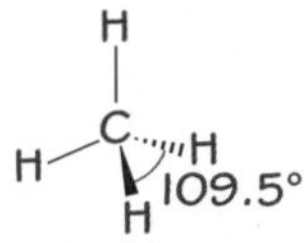 H—C<...H H H 109.5°	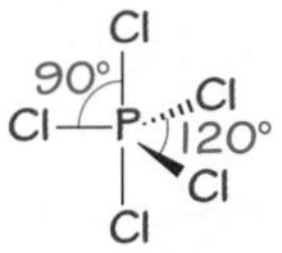 Cl 90° Cl—P...Cl 120° Cl	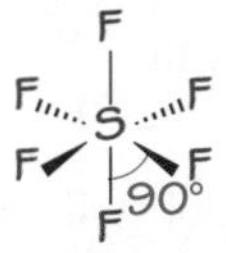 F....S....F F|F 90°

- -

Worked example

1 Determine the shape and bond angle in $BeCl_2$. **(2 marks)**

Shape is linear and the bond angle is 180°.

We can see that Be has 2 electrons in the outer shell and each Cl forms one bond to the Be atom. This gives a total of 4 electrons so 2 electron pairs.

2 Determine the shape of and bond angle in NH_4^+. **(2 marks)**

Shape is tetrahedral, angle is 109.5°.

N has 5 electrons in its outer shell and each H forms a bond to the N, hence 9 electrons. Subtract 1 electron as the ion has a +1 charge, giving a total of 8 electrons so 4 pairs.

Now try this

1 Draw a dot-and-cross diagram of SO_2 and use it to determine the shape of the molecule. **(2 marks)**
2 Predict the shape of and bond angle in the molecule SiH_4. **(2 marks)**

21

More shapes of molecules and ions

Electron pair repulsion theory can also be used to determine the shape of and bond angle in molecules with lone pairs of electrons around their central atoms.

Shapes of molecules with lone pairs of electrons

An orbital containing a lone pair of electrons has a greater electron density. As a result it has a greater ability to repel than a bond pair. These molecules show how this phenomenon affects the bond angle around the central atom.

Be careful! Each lone pair reduces a bond angle by approximately 2.5°.

Dot-and-cross diagram shows 3 bond pairs and 1 lone pair, so the angle is based on a tetrahedron.

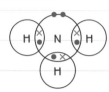

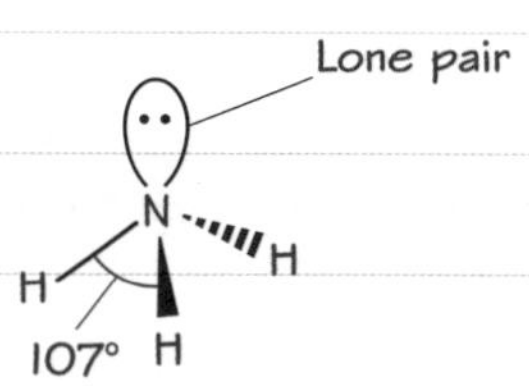

Lone pair

107°

As the lone pair repels more than the bond pairs it reduces the bond angle by about 2.5°. The shape occupied by the four atoms is **pyramidal**.

Dot-and-cross diagram shows 2 bond pairs and 2 lone pairs, so the angle is based on a tetrahedron.

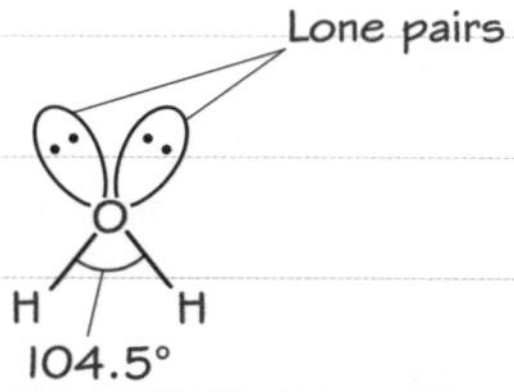

Lone pairs

104.5°

As the lone pairs repel more than the bond pairs they reduce the bond angle by about 5°. The shape occupied by the three atoms is **non-linear**.

Worked example

1 Determine the shape of and bond angle in the molecule PF_3.   **(2 marks)**

Pyramidal, with a bond angle of about 107°.

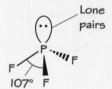
Lone pairs

F   F
107° F

P has 5 electrons in its outer shell and each F forms a bond to the P. This gives a total of 8 electrons so 4 pairs, so angle is based on a tetrahedron.

As the lone pair repels more than the bond pairs it reduces the bond angle by about 2.5°. The shape occupied by the four atoms is **pyramidal**.

2 Determine the shape of and bond angle in the NH_2^- ion.   **(2 marks)**

Non-linear, with a bond angle of about 104.5°.

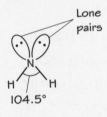
Lone pairs

H   H
104.5°

N has 5 electrons in its outer shell and each H forms a bond to the N, hence 7 electrons. Add one electron as the ion is −1, giving a total of 8 electrons so 4 pairs, so angle is based on a tetrahedron.

As the lone pairs repel more than the bond pairs they reduce the bond angle by about 5°. The shape occupied by the three atoms is **non-linear**.

Now try this

Determine the shape of, and bond angle in, the AlH_4^- ion.   **(2 marks)**

Electronegativity and bond polarity

Electronegativity measures the tendency of an atom to attract a pair of electrons in a covalent bond.

Patterns in electronegativity

Electronegativity can be measured using the Pauling scale. The larger the value the stronger the attraction of the shared pair of electrons to the bonded atom in the covalent bond.

Period → Electronegativity increases →

1	H 2.20																He
2	Li 0.96	Be 1.57											B 2.04	C 2.55	N 3.04	O 3.44	F 2.98	Ne
3	Na 0.93	Mg 1.31											Al 1.61	Si 1.90	P 2.19	S 2.56	Cl 3.16	Ar
4	K 0.82	Ca 1.00	Sc 1.36	Tl 1.54	V 1.63	Cr 1.65	Mn 1.55	Fe 1.83	Co 1.86	Ni 1.91	Cu 1.90	Zn 1.65	Ga 1.81	Ge 2.01	As 2.18	Se 2.55	Br 2.96	Kr 3.00
5	Rb 0.82	Sr 0.95	Y 1.22	Zx 1.33	Nb 1.6	Mo 2.16	Tc 1.9	Ru 2.2	Rh 2.28	Pd 2.20	Ag 1.93	Cd 1.69	In 1.78	Sn 1.96	Sb 2.05	Te 2.1	I 2.66	Xe 2.6

Non-polar bonds

The electronegativity of the atoms in a bond can affect the properties of the

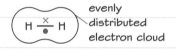

bond. For instance in a H_2 molecule both atoms are identical so have the same electronegativity.

The bond pair of electrons is distributed evenly between the two hydrogen atoms. This type of bond is called a **non-polar** bond. Another similar example is a bond between a carbon atom and a hydrogen atom.

Polar bonds

In HCl the Cl atom has a greater electronegativity than the H atom.

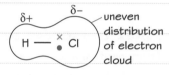

The bond pair of electrons is distributed unevenly as the Cl atom has a stronger attraction for the bond pair than the H atom. This means the chlorine end of the molecule has a partial negative charge ($\delta-$) and, the hydrogen end has a partial positive charge ($\delta+$). This type of bond is called a **polar** bond. The uneven distribution of electrons across the molecule means the molecule is described as having an overall **dipole**.

Suggest how you could test a liquid compound to see if it had an overall dipole.   **(2 marks)**

Run a small stream of the liquid from a burette. If you bring a charged rod towards the liquid the stream will deflect if it is polar.

Will CHF_3 have a dipole moment? Explain your answer.   **(2 marks)**

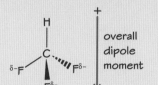

Fluorine is more electronegative than carbon so the C–F bonds will be polar, as shown. The bonds in the molecule are not arranged symmetrically so there is an overall dipole.

Dipole moments

Molecules can have atoms with different electronegativities but not have a dipole moment. For instance, look at CCl_4.

The C–Cl bonds are polar but arranged symmetrically so the dipoles on each bond cancel each other out. Remember, to have an overall dipole moment a molecule must have at least one polar bond and the bonds must not be arranged symmetrically.

Would you expect BCl_3 to be polar? Explain your answer.   **(2 marks)**

Van der Waals' forces

Van der Waals' forces refer to interactions between molecules caused by either permanent or induced dipoles, often referred to as *intermolecular forces*.

Permanent dipole–dipole interactions

Polar molecules such as HCl have permanent dipoles due to the much greater electronegativity of the chlorine atom. Hence the oppositely charged ends of two molecules are attracted to each other. This weak attractive force is called a permanent dipole-dipole interaction.

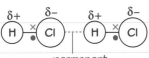

permanent
dipole–dipole interaction

London forces

These forces exist between all simple molecules.

Molecule 1

Electrons are randomly moving within the electron shells of a molecule.

This can cause electron density to be spread unevenly, causing an instantaneous dipole.

Molecule 2

This weak attraction is known as an induced dipole–dipole interaction. It is also sometimes called a London (dispersion) force.

If a second molecule approaches, a small induced dipole occurs, causing a weak attraction between the two molecules.

Intermolecular forces

Permanent dipole–dipole interactions and London forces are examples of intermolecular forces and can be used to explain properties in simple molecules such as boiling point and solubility.

Both are known as van der Waals' forces, though watch out, as some websites and text books may only be referring to London forces when they use this terminology.

London forces can explain the differences in boiling points of the noble gases.

Noble gas	Number of electrons	Boiling point (°C)
He	2	−269
Ne	10	−246
Ar	18	−186
Kr	36	−153
Xe	54	−108
Rn	86	−62

- As you go down the group, the number of electrons per atom increases.
- This increases the strength of the London forces.
- So more energy has to be supplied to overcome the London forces and allow the substance to boil.

Worked example

Describe and explain the trend in boiling points in Group 7.
(3 marks)

As you go down the group the boiling point increases. For instance the boiling point of fluorine is −188°C whereas that of iodine is 184°C. This is because as the molecules get larger they have a greater relative molecular mass, which means more electrons. An iodine molecule has 106 electrons compared to 18 in a fluorine molecule. This increases the strength of the London forces, so more energy has to be supplied to boil the iodine.

Now try this

Put these alkanes in order of boiling point, starting with the lowest. Explain the order.

Butane (C_4H_{10}), ethane (C_2H_6), methane (CH_4), propane (C_3H_8).

(3 marks)

Hydrogen bonding and the properties of water

Water has unusual properties that are explained by an especially strong type of dipole–dipole interaction called a *hydrogen bond*.

Hydrogen bonds

This type of intermolecular force is the attraction between an electron-deficient hydrogen atom ($\delta+$) and a lone pair on oxygen, nitrogen or fluorine atoms. O, N and F are the only atoms that can form hydrogen bonds as they are small and highly electronegative.

Water molecules can form hydrogen bonds between each other.

Hydrogen bonds are an especially strong intermolecular force.

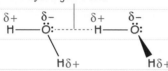

hydrogen bond

Remember

✓ Show the dipole charges on relevant atoms.

✓ Show the lone pairs of electrons on O, N or F.

✓ Indicate the hydrogen bond clearly, e.g. by using a dashed line.

✓ The OHO bond angle should be 180°.

Only electron-deficient hydrogens can form hydrogen bonds, so H atoms attached directly to a carbon **cannot** form hydrogen bonds.

Melting and boiling points of water

The melting and boiling points of water are unusually high for such a small molecule. For instance, its predicted boiling point, based only on London forces, can be estimated by following the trend in the boiling points of Group 6 hydrides is about –100°C.

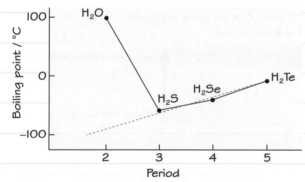

The actual boiling point of water is 100°C. This much higher than expected value is due to the fact that water molecules are able to hydrogen bond to each other.

Density of ice

Ice is an unusual solid as it is less dense than the liquid from which it is formed (water).

This is because the hydrogen bonds in the open lattice keep the molecules apart to a greater extent than in water.

As the ice forms, some of these hydrogen bonds break, allowing the water molecules to pack more closely to each other. This increases the density.

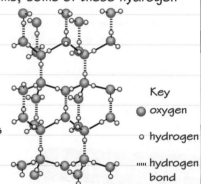

Key
● oxygen
○ hydrogen
⸴⸴⸴ hydrogen bond

Notice that all relevant dipoles are clearly shown and the hydrogen bond is marked as a dashed line. This is to make sure it's not confused with the covalent bonds.

None of the hydrogen atoms attached to the carbon are polar so cannot form hydrogen bonds.

Worked example

Draw a diagram to show a hydrogen bond between two molecules of methanol. **(2 marks)**

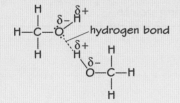

Now try this

The compound HF is a liquid at 10°C but HCl is a gas. Explain this observation in terms of intermolecular forces. **(3 marks)**

Properties of simple molecules

The structure and bonding in simple covalent molecules can be used to explain many of their physical properties.

Simple molecular lattices

Many covalently bonded molecules form simple molecular lattices. These structures have strong covalent bonds between the atoms within the molecule. However the forces between the molecules are weak intermolecular forces, for instance London forces.

An example of such a structure is iodine, a grey-black solid at room temperature.

strong covalent bonds within I_2 molecules

weak London forces between I_2 molecules

Properties

Molecules with simple molecular structures tend to:

 have low melting and boiling points as the intermolecular forces between molecules are weak, so a relatively small amount of energy is needed to cause a change of state

 not conduct electricity as the structures have no free electrons or free ions

 dissolve in solvents with intermolecular forces of similar strength. For instance iodine is soluble in cyclohexane, a solvent with similar London forces.

Covalent bond strength

All covalent bonds are strong but not all bonds are the same strength. Even the same type of bond can have different strengths depending on the environment of the bond. For instance an oxygen-hydrogen bond in water is not quite the same strength as an oxygen-hydrogen bond in ethanol.

Average bond enthalpy

The average amount of energy needed to break one mole of a bond is called the **average bond enthalpy** and this can be used to compare the strength of covalent bonds, as the bigger the value the stronger the bond.

Quoted values from data books have a positive sign, as breaking a bond is endothermic.

Bond	Average bond enthalpy / kJ mol^{-1}
O–H	+464
C–H	+413
C–Cl	+243

↑ Increasing bond enthalpy

↓ Increasing bond strength

Worked example

Explain what happens to the structure of a simple molecular solid when it is heated to its melting temperature.   **(2 marks)**

The energy is sufficient to disrupt the weak intermolecular forces, so the compound melts. However, the strong covalent bonds do **not** break, so the molecules remain intact.

Describing forces and bonds

When justifying melting points it's easy to use incorrect terminology when describing the forces or bonds broken during melting. Forces and bonds need to be used in the correct context to be given credit.

For instance, in simple molecules or simple molecular lattices **weak** intermolecular forces are broken, or disrupted during melting, hence lower melting points.

However, in giant structures, such as giant ionic lattices, many **strong bonds** are broken during melting, hence higher melting points.

Now try this

Ice is an example of a simple molecular solid. Explain why it is a poor conductor of electricity.   **(1 mark)**

Exam skills 2

This exam-style question uses knowledge and skills you have already revised. Have a look at pages 17–19 for a reminder about **electrons**, page 20 for a reminder about **covalent bonds** and pages 24 and 25 on **intermolecular forces**.

Worked example

Nitrogen trifluoride, NF_3, is used in the manufacture of electrical components.

(a) Draw a dot-and-cross diagram to show the bonding in NF_3.   **(1 mark)**

(b) Nitrogen trifluoride will not react with water. However, a similar compound, NH_3 (ammonia), does react with water.

$NF_3 + H_2O \rightarrow$ no reaction

$NH_3 + H_2O \rightarrow NH_4^+ + OH^-$

   (i) Explain how a hydrogen ion bonds to the nitrogen atom in NH_4^+.   **(2 marks)**

The lone pair of electrons from the nitrogen atom forms a dative covalent bond to the hydrogen ion.

> An alternative way to name this bond is to call it a **coordinate** bond.

   (ii) Use the data below and your knowledge of bonding to suggest why NF_3 does not react with water in a similar way to NH_3.

Element	Pauling electronegativity
Nitrogen	3.04
Hydrogen	2.20
Fluorine	3.98

(3 marks)

> 'Suggest' means that you may not have been taught this idea directly, but you should apply your knowledge to an unknown situation. In this case you would have to use your knowledge of dative covalent bonding and electronegativity to suggest why the lone pair on N in NF_3 seems to be unavailable for such bonding.

The electrons from nitrogen are not attracted to the hydrogen atom in ammonia as hydrogen has a lower electronegativity than nitrogen. This means that the lone pair on the nitrogen atom can form a dative covalent bond to a hydrogen ion from water.

Fluorine has a greater electronegativity than nitrogen. This means the three fluorine atoms withdraw electron density from the nitrogen atom. Hence the nitrogen lone pair is pulled closer into the nitrogen atom, so cannot form a dative covalent bond.

(c) The boiling points of NF_3 and NCl_3 are shown in the table. Use your knowledge of intermolecular forces to explain the differences.

Compound	Boiling point / K
NF_3	144
NCl_3	344

(3 marks)

> Don't let your preconceptions stop you from analysing data given in a question. Most students would suggest that greater permanent dipole–dipole interactions are more significant than London forces and in many cases this is true. However, the data shows that it is not so in this case, as the greater London forces outweigh the greater permanent dipole–dipole interactions.

NF_3 is likely to have stronger permanent dipole–dipole interactions as fluorine is more electronegative than chlorine. However, the boiling point of NCl_3 is greater. This is because it has more electrons than NF_3, so has stronger London forces.

The Periodic Table

The **Periodic Table** is an arrangement of the **elements** in order of increasing **atomic number**. Elements with similar chemical properties are found in the same **group**.

The Periodic Table

1	H 2.20																	He
2	Li 0.96	Be 1.57											B 2.04	C 2.55	N 3.04	O 3.44	F 2.98	Ne
3	Na 0.93	Mg 1.31											Al 1.61	Si 1.90	P 2.19	S 2.56	Cl 3.16	Ar
4	K 0.82	Ca 1.00	Sc 1.36	Tl 1.54	V 1.63	Cr 1.65	Mn 1.55	Fe 1.83	Co 1.86	Ni 1.91	Cu 1.90	Zn 1.65	Ga 1.81	Ge 2.01	As 2.18	Se 2.55	Br 2.96	Kr 3.00
5	Rb 0.82	Sr 0.95	Y 1.22	Zx 1.33	Nb 1.6	Mo 2.16	Tc 1.9	Ru 2.2	Rh 2.28	Pd 2.20	Ag 1.93	Cd 1.69	In 1.78	Sn 1.96	Sb 2.05	Te 2.1	I 2.66	Xe 2.6
6	Cs 0.79	Ba 0.89	*	Hf 1.3	Ta 1.5	W 2.36	Re 1.9	Os 2.2	Ir 2.20	Pt 2.28	Au 2.54	Hg 2.00	Tl 1.62	Pb 2.33	Bi 2.02	Po 2.0	At 2.2	Rn
7	Fr 0.7	Ra 0.9	**	Rf	Db	Sg	Bh	Hs	Mt	Ds	Rg	Uub						

s block — across the top row (H ... He)
d block — middle section
p block — right section

A Periodic Table will be on a data sheet in the exams, and is given on page 151 at the back of this book.

Chemical properties of the elements have repeating trends.

1 The reactions of the Period 3 elements with water are shown on page 32.

2 When elements form oxides, the oxides go from basic to acidic across the period.

🖩 Maths skills — Trends in melting points

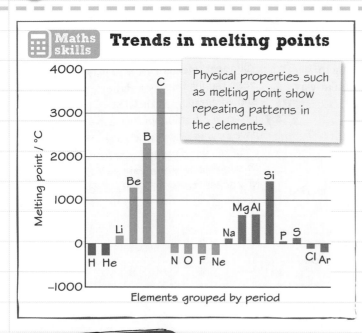

Physical properties such as melting point show repeating patterns in the elements.

(Bar chart: Melting point / °C (vertical axis, −1000 to 4000) against Elements grouped by period (horizontal axis). Elements shown: H, He, Li, Be, B, C, N, O, F, Ne, Na, Mg, Al, Si, P, S, Cl, Ar. Carbon (C) has the highest bar (~3500), Silicon (Si) ~1400.)

Trends in electronic configurations

- Successive elements have an extra electron in their atom.
- In Period 2 the electrons fill the 2s then the 2p orbitals.
- In Period 3 the electrons fill the 3s then the 3p orbitals.
- The block in which an element is found is linked to the orbital/sub-shell of the highest energy electron.

Explanation for the trend in melting point

1 The metals have an increased strength of metallic bonding as the cations get smaller with a higher charge, and there are more delocalised electrons.

2 Carbon and silicon have giant covalent structures with strong covalent bonds throughout.

3 The remaining elements are atoms or molecules with weak intermolecular forces, as described on page 31.

Describe the trend in the melting point of the elements in Periods 2 and 3, stating which element has the highest melting point in each period. **(4 marks)**

The melting point increases for the first four elements in each period. The final four elements in each period have lower melting points, with the Group 0 elements having the lowest melting point. The elements with the highest melting points are carbon in Period 2 and silicon in Period 3.

You need to explain the block, so state which block nitrogen is found in and related to the electronic configuration of a nitrogen atom.

1 Explain why nitrogen is found in the p block. **(1 mark)**

2 Explain the meaning of the term **periodicity**, and give an example using the elements in Period 3. **(3 marks)**

Ionisation energy

The **ionisation energy** is the energy required to remove an outer electron from the attractive force of the nucleus.

An aluminium atom and its first five ionisation energies

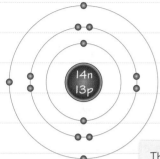

Electron being removed	Ionisation energy / kJ mol^{-1}
1st	580
2nd	1815
3rd	2745
4th	11575
5th	14840

There is a large increase in the energy required to remove the fourth electron. This electron is from an inner shell, closer to the nucleus and with less shielding.

The electronic configuration of an aluminium atom is $1s^2 2s^2 2p^6 3s^2 3p^1$.

First ionisation energies of alkali metals

The **first ionisation energy** is the energy required to remove an electron from each of one mole of gaseous atoms.

$M(g) \rightarrow M^+(g) + e^-$

The table on the right shows the first ionisation energies of the alkali metals.

Further down the group, the outer electron is further from the nucleus and **shielded** by electrons in more inner shells. The electrostatic attraction to the nucleus is reduced.

Element	First Ionisation energy / kJ mol^{-1}
Lithium	520
Sodium	495
Potassium	420

The ionisation energy decreases down the group.

Worked example

Write an equation for the first ionisation energy of chlorine. **(3 marks)**

$Cl(g) \rightarrow Cl^+(g) + e^-$

Although chlorine exists as diatomic molecules, the definition of ionisation energy is for the ionisation of atoms. The (g) state symbol is required.

Explain whether chlorine or bromine has the lower first ionisation energy. **(4 marks)**

Bromine atoms are larger, so the outer electrons are further from the nucleus and shielded by more inner electrons. The electrostatic attraction between bromine's outer electron and the nucleus is less, so the first ionisation energy is lower.

Now try this

The first four ionisation energies of an element are shown in the table.

Electron being removed	Ionisation energy / kJ mol^{-1}
1st	590
2nd	1145
3rd	4910
4th	6490

(a) Plot a graph of the data. **(2 marks)**

(b) Explain what the data indicates about the electronic configuration of the element. **(2 marks)**

(c) Suggest an identity of the element, and explain your answer. **(2 marks)**

Consider using your graph, and looking for a jump in the value of the ionisation energy.

Ionisation energy across Periods 2 and 3

The graph shows the trend in the first ionisation energy of elements in Period 2 and Period 3.

General trend across the period

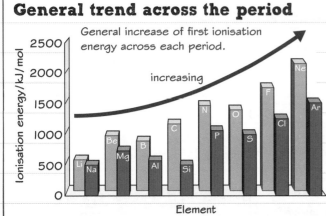

General increase of first ionisation energy across each period.

increasing

Each element in Period 3 has a lower first ionisation energy than the Period 2 element in the same group. This is explained on the previous page.

The first ionisation energy increases.

1 The number of protons in the nucleus increases.

2 The outer electron is in the same shell, with similar shielding.

3 The electrostatic attraction of the outer electron to the nucleus increases.

Worked example

Put the elements carbon, nitrogen and silicon in order of increasing first ionisation energy and explain the order.   **(5 marks)**

The order is silicon, carbon, nitrogen. The outer electrons of carbon and nitrogen atoms are in the same shell, with similar shielding, but nitrogen has the most protons, so greatest electrostatic attraction. Silicon's atom has an extra inner shell, so the outer electron is further from the nucleus and more shielded, giving the least electrostatic attraction.

Exceptions to the general trend

1 Boron and aluminium have lower first ionisation energies than beryllium and magnesium.

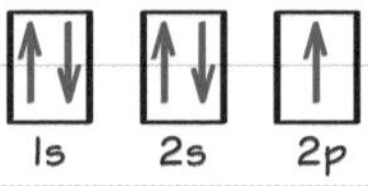

Electronic configuration of boron.

The outer electrons of boron and aluminium are in a p-orbital, higher in energy.

2 Oxygen and sulfur have lower first ionisation energies than nitrogen and phosphorus.

Electronic configuration of oxygen.

The outer electrons of oxygen and sulfur have paired electrons in a p-orbital. These repel each other, so one electron is more easily removed.

Now try this

1 Which element has the highest first ionisation energy?

A Be

B B

C Mg

D Al   **(1 mark)**

2 Which element has the highest second ionisation energy?

A Li

B Be

C B

D C   **(1 mark)**

Structures of the elements

The metallic elements have a **giant metallic lattice** structure. The non-metallic elements have either a **giant covalent lattice** or are composed of **simple molecules** or **atoms**.

The structures of Period 3 elements

Na	Mg	Al	Si	P	S	Cl	Ar
giant metallic lattice			giant covalent lattice	simple molecules			atoms

Giant metallic lattice of sodium

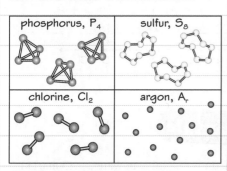

lattice of cations with delocalised electrons

Bonding in metals

There is a strong electrostatic attraction between the cations (positive ions) and the **delocalised electrons** in lithium, beryllium, sodium and magnesium.

Giant covalent lattice of silicon

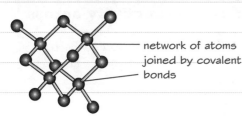

network of atoms joined by covalent bonds

Bonding in carbon and silicon

The covalent bonds between the atoms are strong. These bonds are throughout the structure.

> ### Structures of carbon
>
> There are three different structures of pure carbon – diamond, graphite and graphene.

Molecular structures

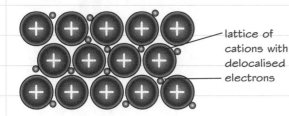

phosphorus, P_4

sulfur, S_8

chlorine, Cl_2

argon, A_r

Molecules with weak intermolecular forces.

Atoms with weak interatomic forces.

Bonding in simple molecules

Nitrogen, oxygen, fluorine, phosphorus, sulfur and chlorine are molecules with strong covalent bonds between the atoms, but weak **intermolecular forces** between the molecules.

Bonding in noble gases

Neon and argon consist of atoms. There are weak interatomic forces between the atoms.

Worked example

State and explain the order of increasing boiling point of the elements phosphorus, sulfur, chlorine and argon. **(3 marks)**

Order is argon < chlorine < phosphorus < sulfur. Argon consists of atoms, chlorine of diatomic molecules, phosphorus of P_4 molecules and sulfur of S_8 molecules. The larger the particles, the stronger the intermolecular forces, and the higher the boiling point.

Now try this

The energy required to break one mole of Si–Si bonds is 222 kJ and for Cl–Cl bonds 240 kJ.

(a) Write an equation showing chlorine melting. **(1 mark)**

(b) Explain why silicon melts at over 1400 °C, but chlorine at the much lower temperature of less than –100 °C. **(3 marks)**

What has to be done in order to change the solid form of these two elements into a liquid? See page 24 for a description of London forces.

Properties of the elements

The properties of the elements vary across the period as their structures change.

Explaining boiling point

1. Metals have high boiling points due to the strong electrostatic attraction between the cations and the delocalised electrons.

2. Carbon and silicon have high boiling points due to the strong covalent bonds that have to be broken.

3. The other non-metals have low boiling points due to weak intermolecular forces.

A graph showing the trend in melting point with an explanation is given on page 28.

Boiling point across Periods 1–3

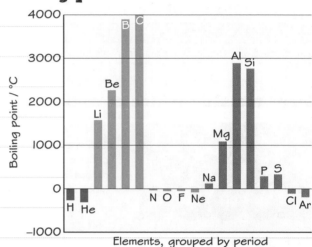

Explaining conductivity

A sample will conduct electricity if there are mobile, charged particles to carry the current.

1. The metal structure has delocalised electrons that can move. Metals are good conductors.

2. Graphite and graphene also have delocalised electrons.

3. Silicon is a **metalloid** and although it has a giant covalent structure, a limited number of electrons can move through the structure, giving a low conductivity.

4. Molecular covalent and Group 0 elements have no free electrons and do not conduct electricity.

Electrical conductivity across Period 3

Metals are good conductors and non-metals are poor conductors.

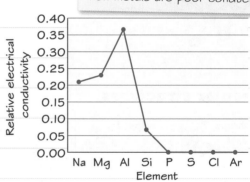

Period 3 elements and water

Element	When cold water is added
Na	vigorous reaction forming NaOH and H_2
Mg	very slow reaction forming $Mg(OH)_2$ and H_2
Al	insoluble
Si	insoluble
P	insoluble
S	insoluble
Cl	soluble (with a slight reaction)
Ar	soluble

Worked example

Suggest why silicon is insoluble in water. **(3 marks)**

Silicon has a giant covalent structure where the silicon atoms are held together by strong covalent bonds. A lot of energy would be needed to break these bonds. Silicon atoms do not interact with water molecules, so no compensating energy would be released.

Now try this

1. Compare, in terms of their structure and bonding, aluminium, silicon and chlorine with respect to
   (a) boiling point **(4 marks)**
   (b) electrical conductivity. **(2 marks)**

Be clear in your answer about the difference between bonding and structure. Bonding refers to how the particles are held together, and can be ionic, covalent or metallic. Structure refers to the assembly of all the particles in a substance, and could be a giant lattice, simple molecules or atoms.

Group 2 elements

The Group 2 elements have atoms with an outer shell electronic configuration of s^2. They form cations with a 2+ charge and a full outer shell, as shown below.

Magnesium atoms are oxidised

$Mg(g) \rightarrow Mg^+(g) + e^-$

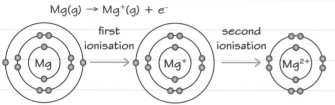

first ionisation → second ionisation

$Mg^+(g) \rightarrow Mg^{2+}(g) + e^-$

> Further down the group, less energy is required to ionise the atoms, as explained on page 29. This is why the reactions of Group 2 elements become more vigorous passing down the group. Redox reactions are described on page 15.

Reactions with oxygen

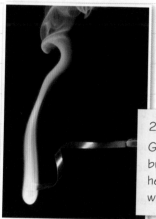

$2Mg + O_2 \rightarrow 2MgO$
Group 2 elements burn vigorously when heated in air to form white oxides.

Reactions of Group 2 elements with water

- Group 2 elements react with water forming a metal hydroxide and hydrogen.
- The general equation, using M for the metal is:
  $M + 2H_2O \rightarrow M(OH)_2 + H_2$

1. Magnesium reacts slightly with cold water (and more vigorously with steam).
2. Calcium reacts to form a white precipitate of calcium hydroxide.
3. Strontium reacts to form less precipitate (with more dissolved product) and barium reacts forming even less precipitate.

Reactions with acids

Group 2 elements react with acids to form a salt and hydrogen.

$Mg + 2HCl \rightarrow MgCl_2 + H_2$

Worked example

Calcium reacts with hydrochloric acid to form calcium chloride.

(a) Write the equation, including state symbols, for this reaction.
(2 marks)

$Ca(s) + 2HCl(aq) \rightarrow CaCl_2(aq) + H_2(g)$

(b) Give an observation that would be made during this reaction.
(1 mark)

Effervescence.

(c) State how the observation would be different if magnesium was used.
(1 mark)

The effervescence would be less vigorous.

Now try this

Magnesium reacts when heated in oxygen.

1 Give **two** observations that could be made when performing this reaction.
(2 marks)

2 Explain, in terms of electrons, why this is a redox reaction.
(2 marks)

> You should consider which atoms gain electrons and which lose electrons.

Group 2 compounds and their uses

Group 2 oxides and carbonates are bases that have uses depending on their ability to neutralise acids.

Reaction of Group 2 oxides with water

Metal oxide	Observation when water is added	Approximate pH of solution formed
MgO	no observable change	8–9
CaO	The white oxide expands	11–12
SrO	as water is added,	12–13
BaO	producing a lot of steam	13–14

The reactions of calcium oxide, strontium oxide and barium oxide are very exothermic.

Solubility

The trend in the solubility of the hydroxides can be used to explain the amount of precipitate formed when the Group 2 elements react with water, shown on the previous page.

Going down Group 2, the hydroxides become more soluble. More OH^- ions are released into the solution, which becomes more alkaline.

$$MO + H_2O \longrightarrow M(OH)_2$$

Uses of Group 2 compounds as bases

Calcium hydroxide is used to neutralise acidic soils.

Calcium carbonate is used to treat acid indigestion.

Magnesium hydroxide is used to treat acid indigestion.

- If soils are too acidic, crops will not grow well. Farmers can add controlled amounts of the base calcium hydroxide which neutralises the acid in the soil to increase the soil pH to the desired value.

- If stomach acid gets into the oesophagus it causes inflammation of the tissues there, causing pain called acid indigestion. If the bases calcium carbonate or magnesium hydroxide are taken they will neutralise this acid, removing the pain.

Worked example

Calcium carbonate reacts with hydrochloric acid.

(a) Write the equation for this reaction. **(2 marks)**

$$CaCO_3 + 2HCl \rightarrow CaCl_2 + H_2O + CO_2$$

(b) Explain how this reaction is used to alleviate indigestion. **(2 marks)**

Indigestion is caused by excess acid in the stomach. Some antacid tablets contain calcium carbonate which neutralises the excess acid.

(c) Suggest why sodium hydroxide tablets are not used to alleviate indigestion. **(2 marks)**

Sodium hydroxide is corrosive and it would cause severe burns in the body.

Now try this

2 g of barium oxide is added to 20 cm³ of water.

(a) Write an equation for the reaction. **(1 mark)**

(b) Suggest the pH of the solution formed. **(1 mark)**

No calculation is expected. The values just give an indication that the concentration of the alkali formed will be quite high.

(c) Explain whether the pH of the solution formed when 2 g of calcium oxide is added to 20 cm³ of water is higher or lower than that in (b). **(3 marks)**

Your explanation should be in terms of the solubility of barium hydroxide and calcium hydroxide.

The halogens and their uses

Chlorine is widely used to treat water, making it fit for drinking.

Boiling points

The halogens, found in Group 7 of the Periodic Table, all exist as diatomic molecules. The boiling point increases down the group.

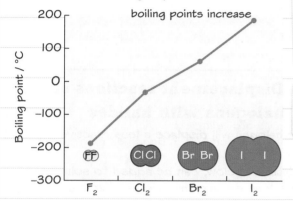

As the size of molecule increases, the induced dipole–dipole interactions (London forces) increase.

A description of London forces is found on page 24.

Uses of chlorine

Bleach is made by reacting chlorine with sodium hydroxide solution to make sodium chlorate(I) solution.

This is an example of **disproportionation**, where the same element, in this case chlorine, is **oxidised** and **reduced** in the same reaction.

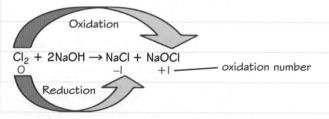

Oxidation

$Cl_2 + 2NaOH \rightarrow NaCl + NaOCl$
  0   -1   +1 ——— oxidation number

Reduction

Use of chlorine in water treatment plants

- ✓ The chlorine kills bacteria in water.
- ✗ Chlorine's toxicity means that care has to be taken in the plant.
- ✗ Chlorine can react with organic matter to form chlorinated hydrocarbons which may be carcinogenic.

Worked example

When chlorine is used to treat water, a chlorine molecule reacts with a water molecule to form chloric(I) acid, HClO.

(a) Write an equation for this equilibrium. **(2 marks)**

$Cl_2 + H_2O \rightleftharpoons HClO + HCl$

(b) Explain why this is classified as a disproportionation reaction. **(5 marks)**

Chlorine's oxidation number changes from 0 in Cl_2 to +1 in HClO, which is oxidation, and to -1 in HCl, which is reduction. The oxidation and reduction of chlorine in this reaction is disproportionation.

When answering a question on disproportionation, you should always evaluate the oxidation numbers of the relevant atoms to show where oxidation and reduction have occurred.

Now try this

Explain why chlorine is used in the treatment of drinking water. Evaluate the advantages and the disadvantages of using chlorine for water treatment. **(4 marks)**

35

Reactivity of the halogens

The halogens, having atoms with an outer shell electronic configuration of s^2p^5, form anions with a 1− charge and a full outer shell, as shown below.

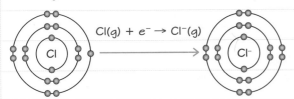

$$Cl(g) + e^- \rightarrow Cl^-(g)$$

Halogen atoms each gain one electron when forming an ionic compound.

Reactivity reduces down the group

Further down the group, less energy is released when the halide ion is formed.

The reactivity reduces because, further down the group:

1. the added electron is further from the nucleus

2. the added electron is shielded by more electrons in inner shells

3. the electrostatic attraction of the added electron to the nucleus is reduced

4. less energy is released by forming the ion.

Cyclohexane has been added to the halogen solutions. The halogens dissolve in the cyclohexane to give a more intense colour. Iodine in water looks brown and not purple, as it does dissolve in cyclohexane.

Practical skills — Displacement reactions of halogens with halides

A more reactive halogen will **displace** a less reactive halogen from its salt.

The halogen solutions, below, can be added to solutions of halide ions, such as sodium chloride, sodium bromide and sodium iodide, which are colourless.

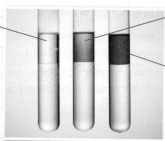

chlorine solution, colourless

bromine solution, orange

iodine solution, purple

Results are:

	Cl⁻ (aq)	Br⁻ (aq)	I⁻ (aq)
Cl_2 (aq)		orange solution	brown solution
Br_2 (aq)	no reaction		brown solution
I_2 (aq)	no reaction	no reaction	

Worked example

Some drops of chlorine solution are added to sodium iodide solution.
(a) Give the observation that would be made. **(1 mark)**

A brown mixture is seen.

(b) Some cyclohexane is added to the mixture formed in (a) and shaken. Explain what is seen. **(2 marks)**

The iodine dissolves into the cyclohexane forming a purple solution with another, aqueous layer.

An ionic equation can be written for each of the reactions occurring.
For example, $Cl_2 + Br_2^- \rightarrow Br_2 + 2Cl^-$
In this reaction chlorine is reduced and bromide ions are oxidised.

Now try this

Describe an experiment that can be carried out in which bromine solution is added to sodium chloride solution and sodium iodide solution. Give the observations, and use these to deduce the relative reactivity of the halogens, chlorine, bromine and iodine, in their redox reactions with halide ions. **(6 marks)**

Split up this question to ensure that you give all of the detail required:
• a description of the test-tube experiments
• the observations made in these experiments
• the equation for the reaction that occurs
• the conclusion (the relative reactivity of the halogens explained from the results).

Tests for ions

 Practical skills Ions can be identified by their reactions to form gases or by precipitation reactions.

Tests for anions: carbonate, sulfate, halides

 Carbonate ions
- Add dilute nitric acid to sample.
- Pass any gas evolved through limewater.

✓ Positive result: white precipitate in limewater. The 'milkiness' in limewater is a suspension of calcium carbonate.

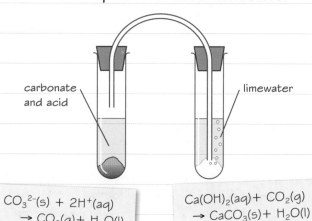

carbonate and acid

limewater

$$CO_3^{2-}(s) + 2H^+(aq) \rightarrow CO_2(g) + H_2O(l)$$

$$Ca(OH)_2(aq) + CO_2(g) \rightarrow CaCO_3(s) + H_2O(l)$$

 Sulfate ions
- Make a solution of the sample.
- Add drops of barium nitrate solution.

✓ Positive result: white precipitate.
$$Ba^{2+}(aq) + SO_4^{2-}(aq) \rightarrow BaSO_4(s)$$
The white precipitate is barium sulfate.

 Halide ions
- Make a solution of the sample.
- Add drops of silver nitrate solution.

✓ Positive result: white precipitate if chloride ions; cream precipitate if bromide ions; yellow precipitate if iodide ions.

- Result confirmed using ammonia.

Test for cation: ammonium

 Ammonium ions
- Add sodium hydroxide solution to sample.
- Warm.
- Test any gas evolved with damp, red litmus paper.

✓ Positive result: litmus paper turns blue.

$$NH_4^+(s) + OH^-(aq) \rightarrow NH_3(g) + H_2O(l)$$

The litmus paper detects alkaline ammonia gas.

Order of tests

If testing an unknown substance for its anion, the tests are carried out in the order 1, 2, 3 given on the left.

✓ Carbonate ions would give a precipitate in tests 2 and 3 if they had not already been detected in test 1. (Alternatively, add dilute nitric acid in tests 2 or 3 which removes carbonate ions.)

✓ Sulfate ions would give a precipitate in test 3 if they have not already been detected in test 2.

$$Ag^+(aq) + Cl^-(aq) \rightarrow AgCl(s)$$
Precipitate dissolves in dilute or concentrated ammonia.

$$Ag^+(aq) + I^-(aq) \rightarrow AgI(s)$$
Precipitate insoluble in dilute and in concentrated ammonia.

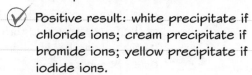

$$Ag^+(aq) + Br^-(aq) \rightarrow AgBr(s)$$
Precipitate dissolves in concentrated ammonia.

Worked example

Given three unlabelled solutions – dilute hydrochloric acid, silver nitrate and sodium iodide – how would you identify which was which using no other substances? **(4 marks)**

Mix all pairs. The pair that gives no precipitate is sodium iodide and hydrochloric acid. The other solution is silver nitrate. Add the silver nitrate to the other solutions. The one that gives a white precipitate is hydrochloric acid, and a yellow precipitate is sodium iodide.

Now try this

1 Write an equation for the reaction of silver nitrate with sodium carbonate. **(2 marks)**
2 Describe what you would see on performing this reaction. **(1 mark)**
3 A solution containing sodium iodide and sodium carbonate has some silver nitrate solution added, followed by some dilute hydrochloric acid. Explain what you would see. **(4 marks)**

Exam skills 3

This exam-style question uses knowledge and skills you have already revised. Look at pages 28–32 and 37 for a reminder about patterns in ionisation energies, the structure of elements in Period 3 and identifying ions with test-tube reactions.

Worked example

(a) The first ionisation energy of some of the elements in Period 3 is shown.

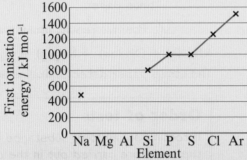

> You should be able to explain the pattern in this graph, including the dips for aluminium and sulfur, which you can see on page 30.

(i) Name, from those elements on the graph, an element from the s-block and an element from the p-block. **(2 marks)**

Sodium is in the s-block and aluminium is in the p-block.

> There are other correct answers for this: magnesium in s-block and silicon to argon in the p-block.
>
> - 'Name' means give the name in word(s).
> - 'Give the formula' requires the chemical formula or symbol.
> - 'Identify' could be answered with a name or formula.

(ii) Explain why the general trend is for the first ionisation energy to increase. **(3 marks)**

The outer electron being removed is in the same shell, so the shielding is similar. However, the number of protons in the nucleus increases across the period. So, the electrostatic attraction between the outer electron and the nucleus increases and more energy is required to remove the electron.

> The answer for this part is shown on the diagram – there is a dip at aluminium.

(iii) Complete the graph to show the missing first ionisation energies. **(2 marks)**

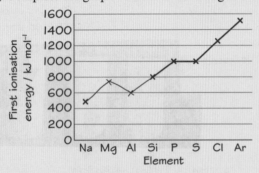

> The boiling points should be explained in terms of structure and bonding. It is worth practising identifying the origin of the forces that hold the particles together in different types of substance.

(b) Give the formulae of the particles present in chlorine gas and argon gas and explain why the boiling point of chlorine is much higher than that of argon. **(4 marks)**

The particles are chlorine molecules, Cl_2, and argon **atoms**, Ar. Chlorine molecules have more electrons than argon atoms, so the London forces are stronger and more energy is needed to overcome them.

(c) Describe a test to show whether some ammonium iodide was contaminating a sample of ammonium carbonate. **(4 marks)**

- Dissolve a sample in distilled water and add nitric acid until the effervescence stops.
- Add silver nitrate solution. If a yellow precipitate forms, add some concentrated ammonia solution. If a yellow precipitate does not dissolve, iodide ions are present.

>
> **Practical skills**   Learn carefully the tests for ions, and the order in which they should be carried out – a carbonate also gives a precipitate with silver nitrate solution. This is why acid is added – it reacts with the carbonate ions giving CO_2. When the effervescence stops, the carbonate ions are all removed and the test can proceed.

Enthalpy profile diagrams

The **enthalpy change**, ΔH, of a reaction is the energy change at constant pressure. In **exothermic** reactions heat is released and in **endothermic** reactions heat is absorbed.

Enthalpy profile diagrams for exothermic and endothermic reactions

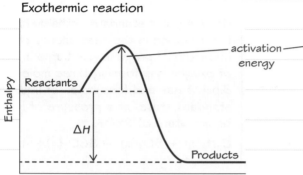 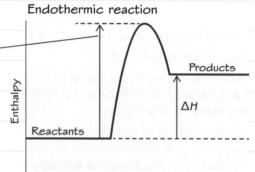

For an exothermic reaction, ΔH is negative. For an endothermic reaction, ΔH is positive.

The activation energy is the minimum energy required for a reaction to take place. Usually, only a small proportion of the particles will have sufficient energy to react.

Worked example

Consider the enthalpy profile diagram on the right.
Identify each of these quantities using: A, B and C, including whether each is negative or positive.

(a) ΔH for the forward reaction.   **(2 marks)**

B, negative

(b) Activation energy for the forward reaction.   **(2 marks)**

C, positive

(c) ΔH for the reverse reaction.   **(2 marks)**

B, positive

(d) Activation energy for the reverse reaction.   **(2 marks)**

B + C, positive

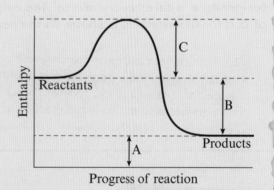

Now try this

Ethanol, C_2H_5OH, fully combusts when ignited to form carbon dioxide and water.

1 Write the equation for this reaction.   **(2 marks)**
2 Draw an enthalpy profile diagram for this reaction, labelling the reactants, products, activation energy and enthalpy change.   **(4 marks)**
3 Explain how the enthalpy profile diagram shows whether this reaction is exothermic or endothermic.   **(2 marks)**
4 Explain why a small beaker containing ethanol can be left in the air, and no reaction occurs, but if lit with a match, the ethanol will burn readily.   **(2 marks)**

Do not forget on your enthalpy profile diagram to label the axes. No values are needed – only the correct shape of the diagram.

Enthalpy change of reaction

The enthalpy change of reaction, $\Delta_r H$, is the heat energy change at constant pressure when the reaction is carried out with the amounts of reactant (in mol) given by the equation.

Standard conditions

Standard conditions are

- pressure 100 kPa
- concentration of solutions $1\,mol\,dm^{-3}$
- at a stated temperature, usually 298 K
- indicated by the symbol $^\ominus$.

The physical state (solid (s), liquid (l) or gas (g)) of a substance at 100 kPa and 298 K is the **standard state** of that substance.

The enthalpy change of reaction under these conditions and with all substances in their standard states is the **standard enthalpy change of reaction**, $\Delta_r H^\ominus$.

Burning propane

When propane is fully combusted the reaction is very exothermic, so water is formed as a gas.

However, the standard enthalpy change of this reaction is the heat energy change when one mole of propane gas burns in five moles of oxygen gas forming three moles of carbon dioxide gas and four moles of **liquid** water (the standard state) at a pressure of 100 kPa and a temperature of 298 K.

$$C_3H_8(g) + 5O_2(g) \rightarrow 3CO_2(g) + 4H_2O(l)$$

 Practical skills   **Measuring the temperature change for the reaction between zinc and copper(II) sulfate solution**

$50\,cm^3$ of $1.0\,mol\,dm^{-3}$ copper sulfate solution is placed in a polystyrene cup. An excess of zinc powder is added. The mixture is stirred and the temperature is taken every minute. The calculation of $\Delta_r H$ for this reaction is shown on the next page.

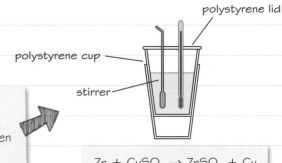
polystyrene lid

polystyrene cup

stirrer

To minimise heat losses, a polystyrene cup with a lid is used. Polystyrene is a poor conductor of heat. A stirrer is used to ensure a quick, complete reaction, and an even temperature distribution in the reaction mixture.

$$Zn + CuSO_4 \rightarrow ZnSO_4 + Cu$$

Worked example

Methane burns in air forming carbon dioxide and water vapour.

(a) Write an equation, including state symbols, for the reaction. **(3 marks)**

$$CH_4(g) + 2O_2(g) \rightarrow CO_2(g) + 2H_2O(g)$$

(b) The enthalpy change of reaction for this reaction is $-810\,kJ\,mol^{-1}$. A data book value for the standard enthalpy change of combustion of methane is $-890\,kJ\,mol^{-1}$. Suggest why the calculated value differs from the data book value. **(2 marks)**

The data book value is for standard conditions. The extra $80\,kJ\,mol^{-1}$ released is due to the exothermic condensation of water.

Now try this

Ammonium chloride dissolves in water. The associated enthalpy change is the **enthalpy change of solution**. Plan a laboratory experiment to measure the temperature change when ammonium chloride is dissolved in water. **(6 marks)**

Do not assume that heat will be released on dissolving – dissolving ammonium chloride in water is endothermic. The method of using the temperature change to calculate the enthalpy change of solution is on the next page.

Calculating enthalpy changes

The enthalpy change can be calculated directly from experimental results if the temperature change can be measured.

Practical skills Using experimental results

The results from the experiment described on page 40 are plotted on the graph below.

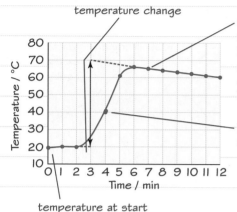

temperature change

temperature falls as reaction is over and heat is lost to the surroundings (this could be a straight line or a curve)

temperature rises as reaction is exothermic

temperature at start

1 The temperature is allowed to stabilise for the first three minutes.

2 The best-fit line is extrapolated backwards to account for heat losses to the surroundings.

Maths skills Calculating enthalpy change from the data

The equation used to calculate the heat released or absorbed is:

$q = mc\Delta T$

- q is the heat change (J)
- m is the mass of water/solution (g)
- c is the specific heat capacity of water ($4.18\,J\,K^{-}\,mol^{-1}$)
- ΔT is the temperature change (K or °C)

In this example,

1 $q = 50 \times 4.18 \times (68 - 20)$

$= 10032\,J = 100.32\,kJ$

The density of water is $1\,g\,cm^{-3}$ so in this example $50\,cm^3$ copper sulfate solution has a mass of $50\,g$

2 The amount of copper sulfate
$= \dfrac{50}{1000} \times 1.0 = 0.050\,mol$

The mass of any solid is not included in the mass, m

3 $\Delta_r H = \dfrac{-10.032}{0.050} = -200.640\,J\,mol^{-1}$

$= -201\,kJ\,mol^{-1}$

The enthalpy change, in $kJ\,mol^{-1}$, is the heat change/mol of reactant. For an exothermic reaction this is negative, and for an endothermic reaction this is positive.

Worked example

A student carried out an experiment to find the enthalpy change when pentanol, $C_5H_{11}OH$, is combusted. $1.00\,g$ of pentanol was burnt, which heated $375\,cm^3$ of water from $20.0\,°C$ to $40.4\,°C$.

(a) Calculate the energy released, in kJ. **(2 marks)**

$q = 375 \times 4.18 \times (40.4 - 20)$

$= 31977\,J$

$= 32.0\,kJ$

(b) Calculate the enthalpy change of the reaction. **(3 marks)**

The amount of pentanol $= \dfrac{1.00}{88.0}$

$= 0.0114\,mol$

$\Delta_r H = -\dfrac{32.0}{0.114} = -2810\,kJ\,mol^{-1}$ (3 s.f.)

Now try this

1 $500\,g$ of glucose, $C_6H_{12}O_6$, was fully combusted. The energy released heated $100\,cm^3$ of water from $22.5\,°C$ to $78.3\,°C$.
Calculate the enthalpy change of the reaction in $kJ\,mol^{-1}$. **(5 marks)**

First calculate the heat change, q, and then the moles of glucose using M_r.

Give the correct sign (here the temperature rises so it is an exothermic reaction), and use an appropriate number of significant figures. The final answer is normally quoted in $kJ\,mol^{-1}$.

Enthalpy change of neutralisation

The enthalpy change of neutralisation, $\Delta_{neut}H$, is the enthalpy change when one mole of water is formed in a neutralisation reaction.

Practical skills — Measuring enthalpy change of neutralisation

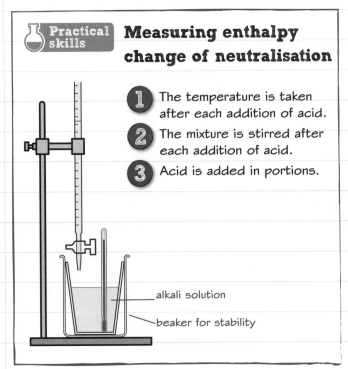

1. The temperature is taken after each addition of acid.
2. The mixture is stirred after each addition of acid.
3. Acid is added in portions.

alkali solution

beaker for stability

Maths skills — Analysing results

A graph is plotted of temperature against volume of acid added.

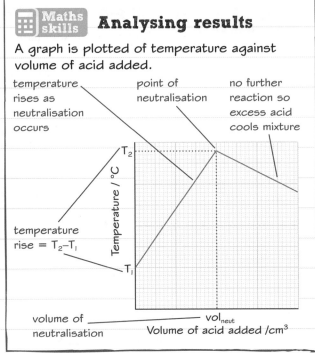

temperature rises as neutralisation occurs

point of neutralisation

no further reaction so excess acid cools mixture

temperature rise = $T_2 - T_1$

volume of neutralisation

vol_{neut}

Volume of acid added /cm^3

Worked example

$150\,cm^3$ of $1.50\,mol\,dm^{-3}$ sodium hydroxide solution is mixed with $150\,cm^3$ of $1.50\,mol\,dm^{-3}$ hydrochloric acid. The initial temperature of the solution is $22.5\,°C$. The final temperature is $31.7\,°C$.

(a) Calculate the enthalpy change of neutralisation, in $kJ\,mol^{-1}$. **(4 marks)**

$q = 300 \times 4.18 \times (31.7 - 22.5)$

$\quad = 11537\,J$

The amount in each solution $= \dfrac{50}{1000} \times 1.50 = 0.225\,mol$

$\Delta_{neut}H = \dfrac{-11537}{0.225} = -51275\,J\,mol^{-1}$

$\quad\quad = -51.3\,kJ\,mol^{-1}$

(b) Suggest two practical ways of minimising heat losses in the experiment. **(2 marks)**

Use a polystyrene container and use a lid.

(c) The data book value for this experiment is $-55.8\,kJ\,mol^{-1}$. Calculate the percentage error in the result. **(4 marks)**

% error $= \dfrac{(55.8 - 51.3)}{55.8} \times 100\% = 8.06\%$

Maths skills

The percentage error is

$\dfrac{\text{true value} - \text{experimental value}}{\text{true value}} \times 100\%$

The ionic equation for neutralisation

The ionic equation for neutralisation is

$H^+ + OH^- \rightarrow H_2O$

For strong acids and alkalis, the other ions are **spectator ions** so do not take part in the reaction. This means that $\Delta_{neut}H$ for a strong acid and base is always $-55.8\,kJ\,mol^{-1}$.

Maths skills

In the calculation the volume of liquid being heated up is $2 \times 150\,cm^3$ ($150\,cm^3$ acid + $150\,cm^3$ alkali)

Now try this

The experiment shown in the diagram and graph above was carried out with sodium hydroxide solution and $25.0\,cm^3$ of $2.00\,mol\,dm^{-3}$ hydrochloric acid. T_1 is $18.4\,°C$ and T_2 is $31.2\,°C$. v_{neut} is $23.4\,cm^3$.

(a) Calculate the energy released in the neutralisation, in kJ. **(2 marks)**

(b) Calculate the enthalpy change of neutralisation, in $kJ\,mol^{-1}$. **(3 marks)**

Hess' law

Hess' law enables the calculation of an enthalpy change when it is not possible to measure the enthalpy change directly. Hess' law is an application of the law of conservation of energy to reactions.

Hess' law

The enthalpy change of a chemical reaction is independent of the route by which the reaction happens.

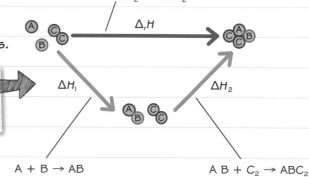

$A + B + C_2 \rightarrow ABC_2$

$\Delta_r H$

The enthalpy change $\Delta_r H$ for the reaction is the same for the different route, $A + B \rightarrow AB$ and then $AB + C_2 \rightarrow ABC_2$

ΔH_1 ΔH_2

$A + B \rightarrow AB$ $A B + C_2 \rightarrow ABC_2$

$$\Delta_r H = \Delta H_1 + \Delta H_2$$

Conditions for Hess' law

For Hess' law to apply, the reactants must start in the same conditions for both routes, and the products must end in the same conditions for both routes.

Forming carbon monoxide directly

The enthalpy change for this reaction cannot be directly measured.

$2C + O_2 \rightarrow 2CO$

- Carbon can be burned in a limited amount of oxygen to form carbon monoxide.
- Carbon dioxide will also form.
- It is not possible to get only carbon monoxide.

Use the cycle on this page.

In ΔH_1 2 moles of CO_2 are formed (2 × first data equation).

In ΔH_2 2 moles of CO are formed (2 × second data equation **reversed**, so +283 is used).

Calculating the enthalpy change for forming carbon monoxide

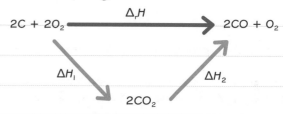

$\Delta_r H$

$2C + 2O_2 \longrightarrow 2CO + O_2$

ΔH_1 ΔH_2

$2CO_2$

$$\Delta_r H = \Delta H_1 + \Delta H_2$$

This **Hess' law cycle** enables us to calculate the enthalpy change for $2C + O_2 \rightarrow 2CO$ (which cannot be measured) if we know the enthalpy change ΔH_1 and the enthalpy change ΔH_2 (which can be measured).

Worked example

Calculate the enthalpy change for $2C + O_2 \rightarrow 2CO$.
(3 marks)

Data:
$C + O_2 \rightarrow CO_2$ $\Delta H = -394\,kJ\,mol^{-1}$
$CO + O_2 \rightarrow CO_2$ $\Delta H = -283\,kJ\,mol^{-1}$

$\Delta_r H = \Delta H_1 + \Delta H_2$

$\Delta H_1 = 2 \times -394 = -788\,kJ\,mol^{-1}$

$\Delta H_2 = 2 \times +283 = +566\,kJ\,mol^{-1}$

$\Delta_r H = -788 + 566 = -222\,kJ\,mol^{-1}$

Now try this

Calculate the enthalpy change for the following reaction:

$N_2(g) + 2O_2(g) \rightarrow 2NO_2(g)$ **(2 marks)**

Data:

$N_2(g) + O_2(g) \rightarrow 2NO(g)$ $\Delta H = +180\,kJ\,mol^{-1}$
$2NO_2(g) \rightarrow 2NO(g) + O_2(g)$ $\Delta H = +112\,kJ\,mol^{-1}$

Draw a Hess' law cycle first to see how the reactions fit together.

Enthalpy change of formation

The enthalpy change of formation, $\Delta_f H$, is the enthalpy change when one mole of a compound is formed from its elements.

Enthalpy change of formation of ethanol, CH_3CH_2OH

As an example, when the following reaction occurs the enthalpy change is the enthalpy change of formation of ethanol.

$$2C(s) + 3H_2(g) + \tfrac{1}{2}O_2(g) \rightarrow CH_3CH_2OH(l)$$

If this reaction is carried out under standard conditions and the reactants and products are in their standard states, then the enthalpy change is the **standard enthalpy change of formation, $\Delta_f H^\ominus$**.

Be careful! The enthalpy change of formation applies when one mole of product is formed. So, for ethanol $\tfrac{1}{2}$ mole of oxygen gas is required and $\tfrac{1}{2}$ must appear in the equation.

 Maths skills ## Calculations using enthalpy of formation data

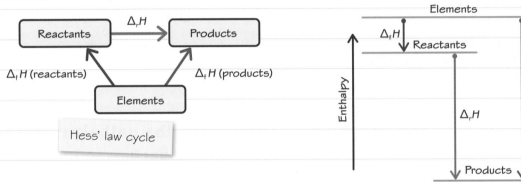

Hess' law cycle

Enthalpy level diagram

$$\Delta_r H = \Sigma\ \Delta_f H\ (products) - \Sigma\ \Delta_f H\ (reactants)$$

Be careful! You can use this formula for any reaction if the **data provided** is enthalpy of formation values. Σ means 'sum', in this case the sum of the $\Delta_f H$ values.

Worked example

Methane reacts with steam to produce carbon monoxide and hydrogen.

$$CH_4 + H_2O \rightarrow CO + 3H_2$$

Compound	$\Delta_f H^\ominus$ / kJ mol^{-1}
CH_4	−75
H_2O	−242
CO	−110

Calculate the standard enthalpy change using the data given. **(3 marks)**

$\Delta_r H = \Sigma\ \Delta_f H\ (products) - \Sigma\ \Delta_f H\ (reactants)$

$\quad = (-110) - (-75 - 242)$

$\quad = (-110) - (-317)$

$\quad = +207\,kJ\,mol^{-1}$

Now try this

PbO reacts with oxygen.

$$3PbO(s) + \tfrac{1}{2}O_2(g) \rightarrow Pb_3O_4(s)$$

Use the $\Delta_f H$ values in table below to calculate the enthalpy change for the reaction. **(3 marks)**

Compound	$\Delta_f H$ / kJ mol^{-1}
PbO(s)	−217
Pb_3O_4(s)	−718

Remember to use the number of moles in the equation. Hydrogen has $\Delta_f H = 0$ by definition because it is an element in its standard state.

Enthalpy change of combustion

The enthalpy change of combustion, Δ_cH, is the enthalpy change when one mole of a substance is completely combusted.

Enthalpy change of combustion of ethanol

As an example, when the following reaction occurs the enthalpy change is the enthalpy change of combustion of ethanol.

$$CH_3CH_2OH(l) + 3O_2(g) \rightarrow 2CO_2(g) + 3H_2O(l)$$

If this reaction is carried out under standard conditions and the reactants and products are in their standard states, then the enthalpy change is the **standard enthalpy change of combustion**, $\Delta_cH^\ominus$.

Conditions

This reaction is exothermic, so water is released as a gas. The standard enthalpy change of combustion requires the reactants and products to be in their standard states, so water has the state symbol (l) in the equation.

Calculations using enthalpy of combustion data

Hess' law cycle

Enthalpy level diagram

$$\Delta_rH = \Sigma \, \Delta_cH \text{ (reactants)} - \Sigma \, \Delta_cH \text{ (products)}$$

Worked example

The table on the right shows some values for standard enthalpy changes of combustion.

Use these values to calculate the standard enthalpy change of the reaction

$$C(s) + 2H_2(g) \rightarrow CH_4(g) \qquad \textbf{(3 marks)}$$

Substance	$\Delta_cH^\ominus$/ kJ mol^{-1}
C(s)	−394
H$_2$(g)	−286
CH$_4$(g)	−890

$$\Delta_rH^\ominus = \Sigma \, \Delta_cH^\ominus \text{ (reactants)} - \Sigma \, \Delta_cH^\ominus \text{ (products)}$$
$$= (-394) + (2 \times -286) - (-890)$$
$$= -966 + 890$$
$$= -76 \, kJ \, mol^{-1}$$

You can use this formula for any reaction if the **data provided** is enthalpy of combustion values. Make sure you know when to use this formula and when to use the formula given on page 44 (to be used when the data provided is enthalpy of formation values – in this formula the order of reactants and products is reversed).

Now try this

The table below gives values for $\Delta_cH^\ominus$ for the first three alkanes.

Alkane	Formula	$\Delta H_c^\ominus$/ kJ mol^{-1}
methane	CH$_4$	−890
ethane	C$_2$H$_6$	−1560
propane	C$_3$H$_8$	−2220

(a) What pattern is shown in this data? **(1 mark)**

(b) Propane can be cracked to form ethene and methane.

$$C_3H_8(g) \rightarrow C_2H_4(g) + CH_4(g)$$
$$\Delta_cH^\ominus \text{ (ethene)} = -1410 \, kJ \, mol^{-1}$$

Use this value together with the values in the table to calculate the enthalpy change of the cracking reaction.

(3 marks)

Bond enthalpies

The **average bond enthalpy** is the enthalpy change, in gaseous molecules, when 1 mole of bonds is broken.

Enthalpy level diagram for $H_2 + Cl_2 \rightarrow 2HCl$

The enthalpy level diagram shows an exothermic reaction – more energy is released forming bonds in the products than was required to break the bonds in the reactants.

In an endothermic reaction, less energy is released forming bonds in the products than was required to break the bonds in the reactants.

A reaction does not necessarily occur in this way (by all the reactants breaking up into atoms and then the atoms forming products). However, Hess' law tells us that the enthalpy change calculated using this model will be valid because the route taken from reactants to products does not matter.

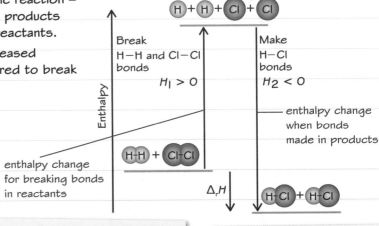

Break
$H-H$ and $Cl-Cl$ bonds
$H_1 > 0$

Make
$H-Cl$ bonds
$H_2 < 0$

enthalpy change when bonds made in products

enthalpy change for breaking bonds in reactants

$\Delta_r H$

bond breaking is endothermic

bond formation is exothermic

Calculating enthalpy changes

The enthalpy change for a reaction:

$\Delta_r H = \Sigma$ bond enthalpies in reactants $- \Sigma$ bond enthalpies in products

Remember!

- It takes more energy to break stronger bonds.
- Shorter bonds are typically stronger.
- Double bonds are stronger than single bonds between the same atoms (see C–C and C=C right), but not twice as strong. See page 62 on the alkenes for the reason.

Examples of bond enthalpies

Bond	Bond enthalpy / $kJ\,mol^{-1}$
H–Br	364
Cl–Cl	239
Br–Br	193
*C–C	347
*C=C	614
*C–H	413
*C–Br	276

*These values are averages across a range of molecules.

Worked example

(a) Calculate the enthalpy change for the reaction

$CH_4 + Br_2 \rightarrow CH_3Br + HBr$ **(3 marks)**

bond enthalpies in reactants $= (4 \times 413) + 193$
$= 1845\,kJ\,mol^{-1}$

bond enthalpies in products
$= (3 \times 413) + 276 + 364$
$= 1879\,kJ\,mol^{-1}$

$\Delta_r H = 1845 - 1879$
$= -34\,kJ\,mol^{-1}$

(b) Explain why the standard enthalpy change of reaction for this reaction given in a data book is different from your answer. **(1 mark)**

The standard enthalpy of formation is an experimental value using the actual substances, but the calculation using bond enthalpies uses average values.

The C–Br bond enthalpy, for example, will be slightly different in different compounds, so the average value used in this calculation may be slightly different from the true value in CH_3Br.

Now try this

1 (a) Write the equation for the reaction of ethene with bromine to make 1,2-dibromoethane. **(1 mark)**

(b) Use the data in the table to calculate the enthalpy change for this reaction. **(2 marks)**

Collision theory

Collision theory is a **model** that explains why changes in the concentration or energy of reactants may alter the **rate of a reaction**.

A mixture of particles undergoing a reaction

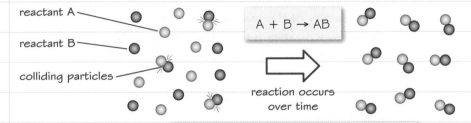

reactant A

reactant B

colliding particles

$A + B \rightarrow AB$

reaction occurs over time

Only a small proportion of the collisions result in a reaction (see page 49).

Collision theory states that for a reaction to occur:

1 the reactant particles must collide

2 the reactant particles must have a minimum amount of energy (the **activation energy**)

3 the reactant particles must collide in the correct orientation.

How to increase the rate of reaction

1 The **frequency of collisions** between the reactant particles can be increased.

2 The energy of the reactant particles can be increased.

3 The reactant particles can be held in the correct orientation.

The ways of achieving these are considered on this and the next few pages.

Increasing the frequency of collisions

The frequency of reactant particle collisions can be increased by having more particles present in the reaction container available to react.

For a reactant in solution, increasing the concentration of the solution increases the number of dissolved particles, and so increases the frequency of collisions.

For a gas reactant, increasing its pressure increases the number of gas particles, and so increases the frequency of collisions.

For a solid reactant, increasing its surface area increases the number of solid particles available to react, and so increases the frequency of collisions.

You have to work out the states of the reactants to tackle this question. There is more than one possible answer.

Worked example

The equations for reactions **A**, **B**, **C** and **D** are

A $CaCO_3 + 2HCl \rightarrow CaCl_2 + H_2O + CO_2$

B $H_2 + Cl_2 \rightarrow 2HCl$

C $Zn + CuSO_4 \rightarrow ZnSO_4 + Cu$

D $Cu + S \rightarrow CuS$

Give a different way to cause each reaction to occur at a faster rate, without a catalyst.   **(4 marks)**

A Increase the temperature of the acid.

B Increase the pressure of the gases.

C Increase the concentration of the copper sulfate solution.

D Powder the solids.

Now try this

Magnesium ribbon reacts with dilute hydrochloric acid. State and explain three ways to increase the rate of reaction.   **(6 marks)**

Do not forget to mention the particles and the frequency of collisions in your answer.

Measuring reaction rates

The **rate of reaction** is the change in the quantity of a reactant or a product divided by the time taken.

Change in quantity with time for the reaction $2A + B \rightarrow 2C$

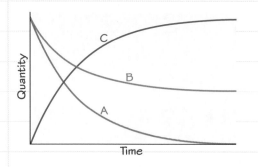

The physical quantity of any reactant or product can be measured. This could be the change in the mass, the concentration of a solution or the volume of a gas.

- The quantity of the reactants falls as the reaction proceeds.
- The **stoichiometric ratio** in the equation shows the relative amounts by which the quantities change.
- The quantity of A drops by twice as much as B.
- The quantity of the product increases as the reaction proceeds.
- The quantity of C increases by the same amount as the quantity of A falls.

Calculating initial rate of reaction using a tangent

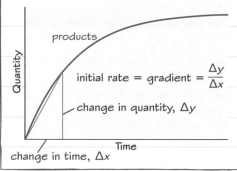

initial rate = gradient = $\dfrac{\Delta y}{\Delta x}$

change in quantity, Δy

change in time, Δx

Calculating average rate of reaction

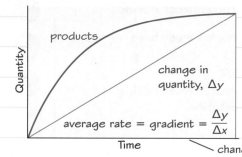

change in quantity, Δy

average rate = gradient = $\dfrac{\Delta y}{\Delta x}$

change in time, Δx

The rate of reaction falls as the reaction proceeds and the reactants are used up.

Worked example

An experiment is carried out to monitor the rate of reaction between calcium carbonate chips and dilute hydrochloric acid. The mass of the reaction vessel is measured.

(a) Explain why the mass of the reaction vessel falls as the reaction occurs. **(1 mark)**

Carbon dioxide leaves the vessel.

Any of the reactants or products can be measured.

(b) Suggest why, when the reaction is complete, the loss of mass may be less than expected. **(1 mark)**

Some of the carbon dioxide dissolves.

(c) Suggest another way of monitoring this reaction. **(1 mark)**

Measure the volume of carbon dioxide using a gas syringe.

Now try this

Hydrogen peroxide, H_2O_2, decomposes to form water and oxygen: $2H_2O_2(aq) \rightarrow 2H_2O(l) + O_2(g)$

A student measures by titration the concentration of hydrogen peroxide solution for a period of time.

Time / min	0.0	0.75	1.5	3.0	5.0	9.0
$[H_2O_2]$ / $mol\,dm^{-3}$	1.0	0.70	0.48	0.18	0.075	0.025

(a) Plot a graph of concentration of hydrogen peroxide solution against time. **(3 marks)**

(b) Use your graph to calculate the rate of reaction after 4 minutes. **(3 marks)**

(c) Suggest another experimental technique to enable you to measure the rate of this reaction. **(2 marks)**

The Boltzmann distribution

The **Boltzmann distribution** shows the distribution of kinetic energy among particles (in a gas).

Only the particles with energy E_a or more will be able to react when they collide. This is only a small proportion of the particles.

Activation energy is defined on page 39.

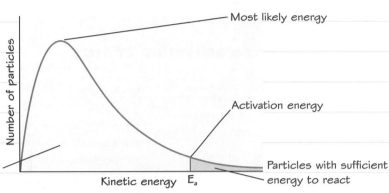

Most likely energy

Activation energy

Particles with insufficient energy to react

Particles with sufficient energy to react

Number of particles

Kinetic energy E_a

The effect of increasing temperature

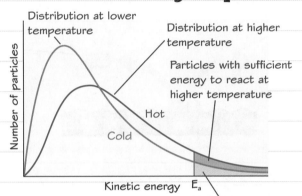

Distribution at lower temperature

Distribution at higher temperature

Particles with sufficient energy to react at higher temperature

Hot

Cold

Number of particles

Kinetic energy E_a

Particles with sufficient energy to react at lower temperature

At a higher temperature a greater proportion of the particles have energy E_A or more and so a greater proportion of collisions will result in a reaction.

Remember!

The particles also collide more frequently at a higher temperature because their average speed is increased. This is a much less important factor than the increased proportion of particles with $E \geq E_A$.

Effect of temperature

At higher temperature, the peak of the curve is further to the right and lower.

Worked example

The diagram shows the Boltzmann distribution for a sample of gas at three temperatures, A, B and C.

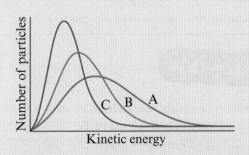

Number of particles

C B A

Kinetic energy

(a) Give the order of temperatures from lowest to highest. **(1 mark)**

C, B, A.

(b) Explain the connection between the areas under the three curves. **(2 marks)**

The area is the same because this equals the total number of particles in the sample.

Now try this

Reaction rates can be altered by changing the temperature of the reaction. The distribution of energy of reactant particles at a specified temperature is shown.

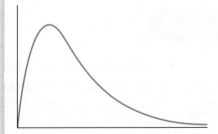

(a) Label the axes on the diagram. **(2 marks)**

(b) Draw a second curve on the diagram to show the distribution of energy of the same reactant particles at a higher temperature. Use this to explain how increasing the temperature increases the rate of reaction. **(4 marks)**

Catalysts

A catalyst is a substance which increases the rate of reaction without being used up by the overall reaction. It allows a reaction to proceed via a different route with lower activation energy. Catalysts are **homogeneous** or **heterogeneous**.

Catalysts reduce the activation energy

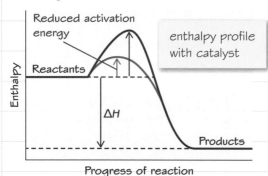

Reduced activation energy

Reactants

enthalpy profile with catalyst

ΔH

Products

Enthalpy

Progress of reaction

More of the particles have $\geq E_{a, \text{CAT}}$ so a greater proportion of collisions result in a reaction.

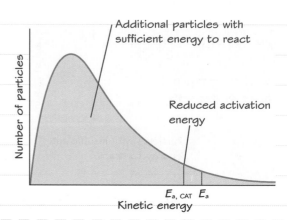

Additional particles with sufficient energy to react

Reduced activation energy

Number of particles

$E_{a, \text{CAT}}$ E_a

Kinetic energy

Heterogeneous catalysts

A heterogeneous catalyst is one in a different **phase** from the reactants. Heterogeneous catalysts are usually solid, with reactants that are gases or liquids.

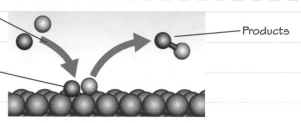

Reactants approach catalyst surface

Products

Reactants adsorbed onto catalyst surface, weakening their bonds

Homogeneous catalysts

A homogeneous catalyst is one in the same phase as (completely mixed with) the reactants. The reaction is split into two steps. Homogeneous catalysts are usually in solution.

Catalyst properties

- Catalysts are not used up, so do not appear in the overall equation.
- Catalysts are often transition metals or their compounds.

Catalysts are used in many industrial reactions

- They lower the temperature required for a suitable rate of reaction.
- At a lower temperature, energy demand is reduced.
- Less fossil fuel needs to be burnt to release energy so less carbon dioxide is released to the atmosphere.
- Manufacturers spend less on energy.

Worked example

Sulfur dioxide reacts with oxygen to form sulfur trioxide.

(a) Write the equation for this reaction. **(1 mark)**

$2SO_2 + O_2 \rightarrow 2SO_3$

(b) Nitrogen monoxide can act as a catalyst for this reaction. In the first step, nitrogen monoxide reacts with oxygen to form nitrogen dioxide. Write the equation for this step. **(1 mark)**

$2NO + O_2 \rightarrow 2NO_2$

(c) Deduce the second step of the catalysed reaction. **(2 marks)**

$SO_2 + NO_2 \rightarrow SO_3 + NO$

Now try this

The reaction between peroxodisulfate ions, $S_2O_8^{2-}$ and iodide ions, I^-, is catalysed by Fe^{3+}.

Step 1: $2I^-(aq) + 2Fe^{3+}(aq) \rightarrow I_2(aq) + 2Fe^{2+}(aq)$

Step 2: $S_2O_8^{2-}(aq) + 2Fe^{2+}(aq) \rightarrow 2SO_4^{2-}(aq) + 2Fe^{3+}(aq)$

1 Write the overall equation. **(1 mark)**

2 Suggest why the catalysed reaction has a lower activation energy. **(3 marks)**

3 Explain whether the catalyst is homogeneous or heterogeneous. **(1 mark)**

The NO_2 formed in step 1 must be used up in step 2 (or the NO would not be a catalyst). Equations for steps 1 and 2 must add up to give the overall equation.

Dynamic equilibrium

When the rate of the **forward reaction** equals the rate of the **backward reaction** in a **closed system**, then a **dynamic equilibrium** exists.

Equilibrium example

This is an example of an equilibrium between hydrogen and iodine (reactants) and hydrogen iodide (product).

Original mixture of gases, hydrogen and iodine.

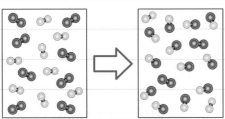

System at equilibrium.

$$H_2(g) + I_2(g) \rightleftharpoons 2HI(g)$$

The mixture is left until there is no further change in concentration of any gas. All three gases are then present and the system is at equilibrium.

Equilibrium

- Hydrogen and iodine gases are mixed in a container.
- The container is sealed to make a closed system (nothing can enter or leave).
- The position of equilibrium can be changed by altering the temperature – in this reaction the change in position can be observed by a colour change (iodine vapour is purple and the other gases are colourless).

Change in concentration as equilibrium is established

When equilibrium is established the forward and reverse reactions continue, but the concentrations of reactants and products no longer change.

- The hydrogen iodide concentration increases by twice as much as the concentration of each reactant falls, as indicated by the stoichiometric ratio in the equation.
- The concentrations of reactant and product are **not** necessarily equal at equilibrium.

Equilibrium established

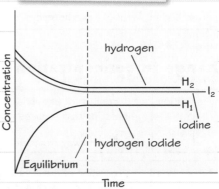

Change in rate as equilibrium is established

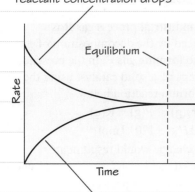

Rate of forward reaction falls as reactant concentration drops

Rate of reverse reaction increases as product concentration rises

When the two rates of reaction become equal, then no further change in concentration can occur. Equilibrium is established.

Which of these could be used to investigate the position of equilibrium at different temperatures for the reaction
$HCOOH(aq) + CH_3OH(aq) \rightleftharpoons HCOOCH_3(aq) + H_2O(l)$
with an acid catalyst?

A a pressure monitor   **B** a thermometer
C titration with alkali   **D** measuring cylinder

C HCOOH is the only acidic substance in the equation.

Now try this

Consider the reaction $H_2(g) + I_2(g) \rightleftharpoons 2HI(g)$
Some hydrogen and iodine are mixed and left until there is no further change in concentration of any substance.

(a) Explain what substances would be found in the equilibrium mixture. **(1 mark)**

(b) Explain why this is described as a **dynamic** equilibrium. **(1 mark)**

Le Chatelier's principle

Le Chatelier's principle states that when any change is made to the conditions of an equilibrium, the position of the equilibrium moves in the direction that minimises the change.

Three positions of equilibrium

$H_2(g) + I_2(g) \rightleftharpoons 2HI(g)$

Position of equilibrium is far to left (mainly reactants).

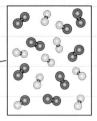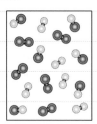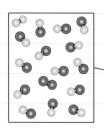

Position of equilibrium is far to right (mainly products).

Change in temperature

temperature ↑	position of equilibrium moves in direction that absorbs heat
	endothermic direction

temperature ↓	position of equilibrium moves in direction that releases heat
	exothermic direction

Change in pressure

pressure ↑	position of equilibrium moves in direction that reduces pressure
	direction giving **fewer** gas molecules

pressure ↓	position of equilibrium moves in direction that increases pressure
	direction giving **more** gas molecules

Change in concentration

concentration ↑	position of equilibrium moves in direction that reduces concentration
	direction **away** from added substance

concentration ↓	position of equilibrium moves in direction that increases concentration
	direction **towards** removed substance

Addition or removal of a catalyst

catalyst added +	position of equilibrium **unchanged**

catalyst removed −	position of equilibrium **unchanged**

A catalyst increases the rate of the forward and backward reactions equally so does not affect the position of equilibrium.

Worked example

Nitrogen reacts with oxygen to form nitrogen monoxide.

$$N_2(g) + O_2(g) \rightleftharpoons 2NO(g) \quad \Delta H = +180 \, kJ \, mol^{-1}$$

(a) Explain the effect on the yield of NO of an increase in temperature. **(3 marks)**

The equilibrium moves in the direction that absorbs heat, the endothermic direction, which is to the right, so the yield increases.

(b) Explain the effect on the yield of NO of an increase in pressure. **(2 marks)**

The position of equilibrium is unchanged as there are 2 moles of gas on each side of the equation, hence no change in yield.

(c) Explain the effect on the yield of NO when adding a liquid that absorbs oxygen. **(2 marks)**

The equilibrium moves in the direction that forms oxygen, which is to the left, so the yield falls.

Now try this

In an industrial process gas **R** is produced by the decomposition of gas **Q**, also forming gas **P**, in the presence of lumps of a solid catalyst using the equilibrium reaction below.

$$Q(g) \rightleftharpoons P(g) + R(g)$$
$$\Delta H = +120 \, kJ \, mol^{-1}$$

What change would result in an increased yield of **R**? **(1 mark)**

A Increasing the pressure

B Increasing the temperature

C Making the catalyst into a powder

D Changing the catalyst to a homogeneous one

The equilibrium constant

The equilibrium constant, K_c is the ratio of product concentration to reactant concentrations, raised to the appropriate power. This only varies for a reaction when the temperature is changed, and it indicates whether the position of equilibrium is to the left or right. (See module 5 for finding K_c.)

Expression for equilibrium constant

For the reaction $aA + bB \rightleftharpoons cC + dD$

concentration of product (C)

power is the number of moles (of D) in the equation

$$K_c = \frac{[C]^c [D]^d}{[A]^a [B]^b}$$

concentration of reactant (A)

power is the number of moles (of B) in the equation

Equilibrium constant for ammonia synthesis

The reaction is $N_2 + 3H_2 \rightleftharpoons 2NH_3$

concentration of ammonia

power is 2; 2 moles of ammonia in the equation

$$K_c = \frac{[NH_3]^2}{[N_2][H_2]^3}$$

concentration of nitrogen

power is 3; 3 moles of hydrogen in the equation

Conditions used in industry

In industry a compromise may be necessary between yield and rate.

1 For an exothermic reaction, a lower temperature gives a higher yield (see page 52).

2 A lower temperature saves energy costs.

3 However, a lower temperature slows down the forward and backward reactions so it would take longer to reach equilibrium.

A compromise temperature will be used.

Estimating the position of equilibrium

- A high value of K_c ($\gg 1$) indicates a position of equilibrium to the right.
- A low value of K_c ($\ll 1$) indicates a position of equilibrium to the left.

K_c for a reaction applies only once the system has reached equilibrium.

Worked example

Nitrogen and hydrogen are mixed in a container and left until equilibrium is reached, according to the equation above.

The equilibrium mixture contained:

$0.127 \, \text{mol dm}^{-3}$ hydrogen

$0.0402 \, \text{mol dm}^{-3}$ nitrogen

$0.00272 \, \text{mol dm}^{-3}$ ammonia.

Calculate K_c.   **(3 marks)**

$$K_c = \frac{[NH_3]^2}{[N_2][H_2]^3}$$

$$= \frac{(0.00272 \, \text{mol dm}^{-3})^2}{(0.0402 \, \text{mol dm}^{-3})(0.127 \, \text{mol dm}^{-3})^3}$$

$$= \frac{7.3984 \times 10^{-6} \, \text{mol}^2 \, \text{dm}^{-6}}{8.2345 \times 10^{-5} \, \text{mol}^4 \, \text{dm}^{-12}}$$

$$= 0.0898 \, \text{dm}^6 \, \text{mol}^{-2}$$

 Maths skills For indices, the number on the bottom line is subtracted from the number on the top line:

dm^3: $(-6) - (-12) = -6 + 12 = 6$

mol: $2 - 4 = -2$

Now try this

NO(g), H_2(g), N_2(g) and H_2O(g) exist in equilibrium:

$2NO(g) + 2H_2(g) \rightleftharpoons N_2(g) + 2H_2O(g)$

The equilibrium is well to the right-hand side so $K_c \gg 1$.

At room temperature and pressure, the equilibrium lies well to the right hand side.

Which of the following could be the value for the equilibrium constant for this equilibrium? **(1 mark)**

A 1.5×10^{-3} B 0.75   C 1.0   D 6.5×10^2

Exam skills 4

This exam-style question uses knowledge and skills you have already revised. Look at pages 39, 42, 46 and 53 for a reminder about enthalpy profile diagrams, enthalpy of neutralisation, using bond enthalpies and the equilibrium constant.

Worked example

Hydrogen reacts with fluorine to form hydrogen fluoride.

$$H_2 + F_2 \rightleftharpoons 2HF$$

(a) (i) Write an expression for the equilibrium constant, K_c.　**(1 mark)**

$$K_c = \frac{[HF]^2}{[H_2][F_2]}$$

> Remember to include the numbers in the equation as powers in the equilibrium expression.

(ii) At equilibrium, there are 0.25 mol of hydrogen, 0.30 mol of fluorine and 2.0 mol of hydrogen fluoride in a container of volume 10 dm³. Calculate K_c, with units if any.　**(2 marks)**

$$K_c = \frac{\left(\frac{2.0}{10}\right)^2}{\left(\frac{0.25}{10}\right)\left(\frac{0.30}{10}\right)} = \frac{4}{0.075} = 53 \text{ (no units)}$$

> **Maths skills** Square brackets mean concentration in $mol\,dm^{-3}$. When substituting into K_c, the moles are divided by the volume (10 dm³) to get concentration. In this case, there are the same number of moles on each side of the equation, so the volumes cancel. It is still useful to include the volume so that it is not forgotten when it does not cancel. This also means that K_c has no units in this case.

(b) (i) Calculate the standard enthalpy change of this reaction, given the average bond enthalpies below.　**(3 marks)**

Bond	Average bond enthalpy / kJ mol⁻¹
H–H	432
F–F	159
H–F	569

Enthalpy of bonds broken = 432 + 159 = 591 kJ mol⁻¹
Enthalpy of bonds formed = 2 × 569 = 1138 kJ mol⁻¹

$\Delta_r H$ = 591 – 1138
= –547 kJ mol⁻¹

(ii) Using your answer to (b)(i), sketch an energy profile diagram for the reaction, labelling E_A and ΔH.　**(3 marks)**

> **Maths skills** Carefully show your working so that if you make a slip the examiner can follow and award part-marks. Don't forget units.

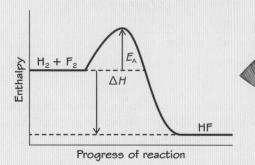

> The shape of this diagram is important – see page 39 for the different enthalpy profile diagrams for exothermic and endothermic reactions. You can see that this diagram is for an exothermic reaction, because the answer to (b)(ii) is negative.

(c) The standard enthalpy of neutralisation of hydrofluoric acid, a weak acid, with sodium hydroxide solution is −59 kJ mol⁻¹. Suggest why this is different from the standard enthalpy of neutralisation of hydrochloric acid with sodium hydroxide solution.　**(3 marks)**

> 'Suggest' means that you may not have been taught this idea directly, but you should apply your knowledge to an unknown situation. In this case, the information that HF is a weak acid is the key to forming an answer.

Strong acids and alkalis are fully ionised, so neutralisation is $H^+ + OH^- \rightarrow H_2O$. As HF is a weak acid, only a small proportion of the molecules are ionised. When undissociated HF molecules are changed into ions in solution, there is an enthalpy change which alters the standard enthalpy of neutralisation.

Key terms in organic chemistry

Lots of key terms not seen at GCSE are used in organic chemistry.

Classifying organic compounds

Features of an organic compound can be used to group it with similar compounds. The most important is its **functional group**. This is a group of atoms responsible for its characteristic reactions.

A **homologous series** is a series of organic compounds with the same functional group, in which each successive member differs by CH_2.

An **alkyl group** is a hydrocarbon group with the formula C_nH_{2n+1}, such as methyl (CH_3). They are often branches, off the main carbon chain.

Aliphatic hydrocarbons have carbon atoms in either straight or branched chains, and those aliphatic hydrocarbons, with some carbon atoms joined in a non-aromatic ring structure, are called **alicyclic**.

cyclohexane C_6H_{12}
alicyclic ring

benzene C_6H_6
aromatic ring

An **aromatic** compound contains a specific type of ring, called a benzene ring.

Saturated hydrocarbons contain single carbon–carbon bonds only, but **unsaturated** hydrocarbons contain at least one carbon–carbon multiple bond.

Key functional groups in Module 4

Type of compound	Formula	Prefix	Suffix
alcohol	–OH	hydroxy-	-ol
aldehyde	–CHO		-al
alkane	C–C		-ane
alkene	C=C		-ene
carboxylic acid	–COOH		-oic acid
halogenoalkane	–F –Cl –Br –I	fluoro- chloro- bromo- iodo-	
ketone	C—CO—C		-one

Bond breaking

The initial stage of a reaction in organic chemistry involves the breaking of a covalent bond. There are two ways in which this can occur.

Homolytic fission

Each electron in the bond pair goes to a different atom in the bond. This can be shown as: $A–B \rightarrow A^{\cdot} + B^{\cdot}$

The species formed are called **radicals** and the dot represents an unpaired electron. Radicals are neutral and very reactive.

Heterolytic fission

Both electrons in the bond pair go to one atom in the bond. This can be shown as:

$A–B \rightarrow A^+ + B^-$

The species formed are positive and negative ions.

Worked example

Classify these compounds according to the definitions on this page.

(a) $CH_3CH_2CH=CH_2$ **(2 marks)**

Aliphatic hydrocarbon, as a straight chain, and unsaturated, as contains a carbon–carbon multiple bond.

(b)

           (2 marks)

Alicyclic ketone, has a non-benzene ring and a ketone functional group.

Now try this

1 Write an equation to show the homolytic fission of a Cl–Cl bond.   **(1 mark)**

2 Suggest how the bond in H–Cl will break and explain your answer.   **(2 marks)**

3 Classify this compound according to the definitions on this page:    **(2 marks)**

You may find it helpful to think about the polarity of the bond.

Naming hydrocarbons

The huge number and variety of organic compounds means that a systematic way of naming them is required. The IUPAC rules ensure that chemists all over the globe have a consistent approach to naming.

Key rules for naming hydrocarbons

- Find the longest unbranched chain of carbon atoms. This determines the **stem** of the name.
- Identify any side chains, often alkyl groups, come from the longest unbranched chain.
  These determine any prefixes in the name, added in front of the stem.
- Identify the key functional group. For hydrocarbons these will be alkane or alkene.
  This determines the suffix in the name, added after the stem.

Worked example

Name the compounds shown.

(a)

(1 mark)

Four carbons in the longest unbranched chain so the stem is **but-**.

There is a methyl group on the second carbon so the prefix is **2-methyl**.

The compound is an alkane so the suffix is **-ane**.

2-methylbutane

(b)

(1 mark)

Five carbons in the longest unbranched chain so the stem is **pent-**.

There are methyl groups on the second and third carbons so the prefix is **2,3-dimethyl**. The 'di' indicates two methyl groups, the numbers show their positions on the unbranched chain.

The compound is an alkane so the suffix is **-ane**.

2,3-dimethylpentane

(c)

(1 mark)

Five carbons in the longest unbranched chain so the stem is **pent-**.

No side chains so no prefix required.

The compound is an alkene so the suffix is **-ene**.

The position of the double bond has to be stated using the lowest numbered carbon in the double bond, in this case -1-.

pent-1-ene

(d)

(1 mark)

Eight carbons in the longest unbranched chain so the stem is **oct-**.

Ethyl group on fourth carbon, methyl group on third. Different side chains go in alphabetical order so prefix is **4-ethyl-3-methyl**.

The compound is an alkane so the suffix is **-ane**.

4-ethyl-3-methyloctane

Now try this

Draw the displayed formulae of these compounds:
(a) 2,2,3-trimethylheptane   **(1 mark)**
(b) 2-methylbut-2-ene   **(1 mark)**
(c) 2,5-dimethylhept-2-ene   **(1 mark)**

Number the chain from the end that gives the smallest numbers possible when naming compounds.
For instance

is 2-methylhexane
not 5-methylhexane

Naming compounds with functional groups

More complex compounds containing a variety of functional groups can be named using the same steps as when naming hydrocarbons.

Key rules for naming compounds with functional groups

- Find the longest unbranched chain of carbon atoms. This determines the stem of the name.
- Identify any side chains coming from the longest unbranched chain. These determine any prefixes in the name, added in front of the stem.
- Identify the key functional group. This determines the suffix in the name, added after the stem.
- Number any functional groups to show their position on the chain.

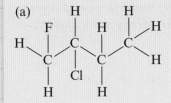

Worked example

Name the compounds shown

(a) 2-Chloro-1-fluorobutane

(1 mark)

Four carbons in the longest unbranched chain so the stem is **but-**.

The compound is a haloalkane so the suffix is **-ane**.

Haloalkanes are named as alkanes with halogen atoms as side groups.

There is a fluorine atom on the first carbon and a chlorine atom on the second carbon. The different halogens go in alphabetical order so the prefix is **2-chloro-1-fluoro**.

(b) Pentan-2-ol

(1 mark)

Five carbons in the longest unbranched chain so the stem is **pent-**.

There are no side groups, so no prefix required.

The compound is an alcohol so the suffix is **-ol**.

The alcohol functional group is on the second carbon so is numbered 2.

(c) Hexan-3-one

(1 mark)

Six carbons in the longest unbranched chain so the stem is **hex-**.

No side chains so no prefix required.

The compound is a ketone so the suffix is **-one**.

The position of the carbon–oxygen double bond has to be stated using the number of the carbon in the double bond, in this case -3-.

(d) 2-methylbutanoic acid

(1 mark)

Four carbons in the longest unbranched chain so the stem is **but-**.

The compound is a carboxylic acid so the suffix is **-oic acid**.

Methyl group on second carbon as position of functional group determines counting order, so prefix is **2-methyl**.

Note in (b), (c) and (d) the suffix begins with a vowel, so the alkane stem has the –e– on the end removed.

Hence pentan-2-ol **not** pentane-2-ol.

Now try this

Draw the displayed formulae of these compounds:
(a) 1,1,1-trichloroethane   **(1 mark)**   (c) 3-chloroheptan-2-ol   **(1 mark)**
(b) 2-methylbutanal   **(1 mark)**   (d) propen-1-ol   **(1 mark)**

Different types of formulae

Different kinds of formulae are used in organic chemistry, each providing differing levels of information about the organic molecule they represent.

Molecular formulae

Molecular formulae simply show the type and number of each atom present in a molecule. This can be determined by calculating the empirical formula, the simplest ratio of atoms, of a substance and comparing it to its relative molecular mass. You need to have, or be able to work out, data on the mass or percentage of each element present in order to do this.

Displayed formulae

These show all the bonds present in a molecule so it is clear to see how each atom joins to the others in the molecule. The diagram shows the displayed formula of ethanoic acid.

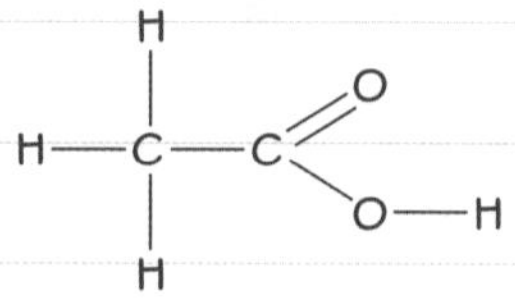

Note that all the bonds are shown, including the bond between the oxygen and hydrogen atoms.

Structural formulae

These show how the atoms are arranged but do not require all the bonds to be shown. For instance, butane has the molecular formula C_4H_{10} and its structural formula is $CH_3CH_2CH_2CH_3$.

Side chains are shown in brackets, to the right of the carbon atoms to which they are attached.

For example, the displayed formula of 2-methylbutane is –

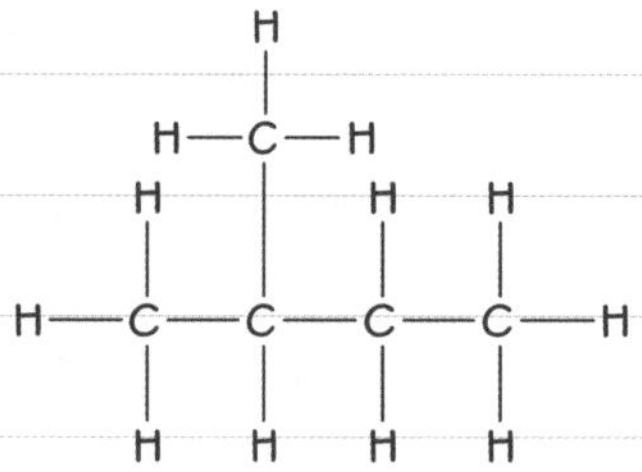

So its structural formula is $CH_3CH(CH_3)CH_2CH_3$.

Skeletal formulae

These are simplified formulae with all the hydrogens attached to the main carbon chain and alkyl groups removed. Once you've practised drawing them, they are easy to use as you don't have to spend ages making sure all the hydrogen atoms are included!

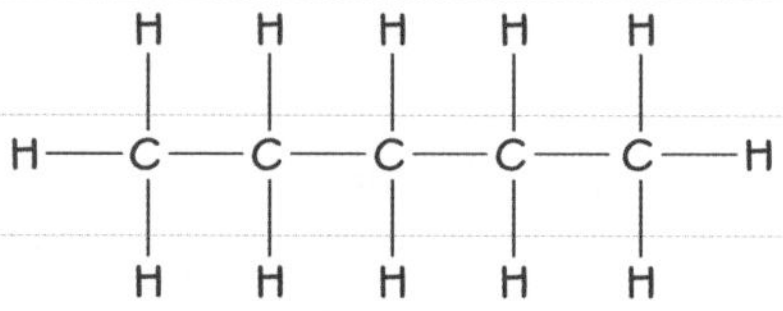

pentane displayed    pentane skeletal

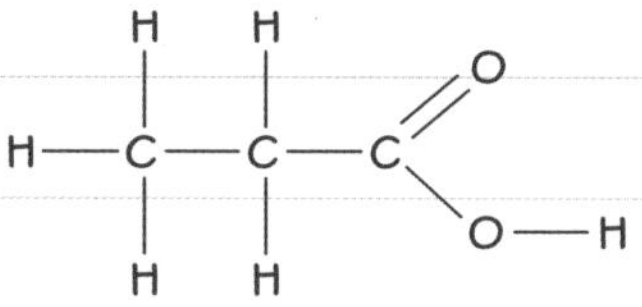

propanoic acid displayed 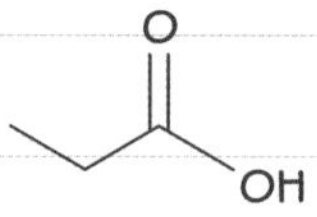   **propanoic** acid skeletal

Worked example

Assuming there is 100 g of the compound the percentages are equivalent to masses in grams. Hence you can find the molar ratio, by calculating the amount of each element, then dividing through by the smallest amount. If this is not a whole number ratio, then scale up as appropriate.

A compound contains 60.00% carbon, 4.44% hydrogen and 35.56% oxygen. Its M_r is 180. Use this information to calculate its molecular formula.

	Carbon	Hydrogen	Oxygen
Amount (mol)	$\dfrac{60.00}{12.0} = 5\,mol$	$\dfrac{4.44}{1.0} = 4.44\,mol$	$\dfrac{35.56}{16.0} = 2.22\,mol$
Molar ratio	$\dfrac{5}{2.22} = 2.25$	$\dfrac{4.44}{2.22} = 2$	$\dfrac{2.22}{2.22} = 1$
Whole number ratio	9	8	4

So, the empirical formula (simplest whole number ratio) is $C_9H_8O_4$. This has a mass of 180.0, so must also be the molecular formula. The molecular forumula is also $C_9H_8O_4$.

Now try this

Draw displayed, structural and skeletal formulae of:

(a) 2-methylpentan-2-ol **(3 marks)**

(b) propanal **(3 marks)**

(c) chloroethanoic acid. **(3 marks)**

Structural isomers

Structural isomers are molecules with the same molecular formula whose atoms are arranged differently and so have a different structural formula.

Isomers due to branching

Often hydrocarbons can have a variety of isomers due to the range of possible branches on a carbon chain. For instance, the molecule C_5H_{12} has three isomers due to branching, with different structural formulae.

pentane

2-methylbutane

dimethylpropane

Be careful! Sometimes the way molecules are presented can make them look like they may be isomers. For instance, in this example, molecule Q could be mistaken for a new isomer of pentane.

'molecule Q'

rotate 180°

However, rotating the molecule shows that it is 2-methylbutane.

Isomers due to the position of the functional group

Functional groups can be positioned at different points in a carbon chain, resulting in different isomers. For instance, the hydroxyl group in an alcohol with the molecular formula C_3H_8O can bond to the carbon chain in two different positions.

OH propan-1-ol

OH

propan-2-ol

Isomers due to different functional groups

With similar functional groups a rearrangement of the atoms can result in compounds with the same molecular formula but with a different functional group. For instance, the molecule C_3H_6O can be arranged in two different ways giving two different functional groups.

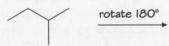

O } aldehyde functional group

propanal

O } ketone functional group

propanone

Worked example

Draw skeletal formulae of the five structural isomers with the formula C_6H_{14}. **(5 marks)**

hexane   2-methylpentane   3-methylpentane   2,3-dimethylbutane   2,2-dimethylbutane

One way to approach this problem is to draw the unbranched isomer first, then branch using a methyl group from left to right on a five-carbon chain to find isomers.

Once all the possible isomers with a single methyl branch have been identified, branch using two methyl groups on a four-carbon chain to find isomers.

It's helpful to name the isomers to check you haven't repeated the same isomer drawn in a different way. If you have access to molecular models when revising, use them to make models of the isomers.

Note there is no 3,3-dimethylbutane. This would be 2,2-dimethylbutane rotated through 180°.

Now try this

Draw skeletal formulae of the nine structural isomers with the formula C_7H_{16}. **(9 marks)**

Properties and reactivity of alkanes

The alkane homologous series consists of saturated hydrocarbons of the general formula C_nH_{2n+2}. Their bonding explains key chemical and physical properties.

Bonding in alkanes

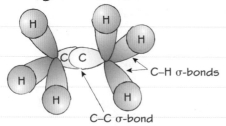

C–H σ-bonds

C–C σ-bond

The overlap of orbitals in alkanes results in the formation of sigma (σ) bonds between carbons and hydrogens, and between adjacent carbons. These single bonds are free to rotate.

Maths skills — Shape of alkanes

The four σ bond pairs around each carbon in alkanes repel equally. This means the shape around each carbon in an alkane is tetrahedral, with a bond angle of 109.5°.

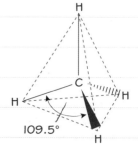

109.5°

Maths skills — Boiling points of alkanes 1

The longer the carbon chain of a linear alkane, the higher the boiling point.

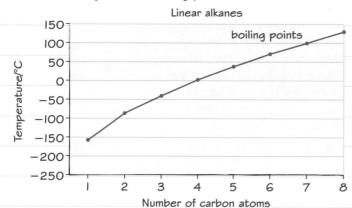

Linear alkanes

boiling points

Number of carbon atoms

Longer carbon chains have stronger London forces as they have more electrons. The electron density of the larger electron clouds fluctuates more readily so instantaneous dipoles are stronger in magnitude. Hence more energy is required to break the London forces.

Boiling points of alkanes 2

The more branching an alkane has, the lower its boiling point.

Alkane	Boiling point / °C
hexane	69
2-methylpentane	60
2,2-dimethylbutane	50

Isomers with more branching have weaker London forces as the branches get in the way of surface area interactions between molecules. Hence less energy needs to be supplied to break the London forces.

Worked example

Explain how the bonding in ethane affects its reactivity.
(3 marks)

Alkanes are generally unreactive.

In the case of ethane both the C–H bonds and C–C bonds have relatively high bond enthalpies so require large amounts of energy to break them.

The electronegativities of C and H are similar so the bonds in ethane are non-polar. This means attacking species such as nucleophiles or electrophiles are not attracted to the atoms in these bonds.

Now try this

1 Explain whether 2-methylhexane or 2,2-dimethylpentane has the higher boiling point. **(2 marks)**

2 Would you expect methane to react with hydroxide ions at room temperature and pressure? Explain your answer. **(2 marks)**

Reactions of alkanes

The most significant reactions of alkanes are combustion and the reaction with halogens such as bromine.

Combustion of alkane fuels

Combustion is the reaction with oxygen.

Vehicles burn hydrocarbon fuels such as petrol or diesel.

Petrol contains hydro-carbons with 4-12 carbons.

Diesel contains hydrocarbons with 8-21 carbons.

Complete combustion produces CO_2 and H_2O. Occurs in excess oxygen.

Incomplete combustion produces CO and C. Occurs in limited oxygen.

When writing equations for combustion reactions it's helpful to follow some basic rules:

 Write the formulae of the reactants and products.

 Balance the carbon atoms first.

 Balance the hydrogen atoms using the water molecules.

 Finally balance the oxygen atoms. As oxygen gas is diatomic, using halves is acceptable.

For example: complete combustion of octane

$$C_8H_{18}(l) + 12\tfrac{1}{2}O_2(g) \rightarrow 8CO_2(g) + 9H_2O(l)$$

Reaction of ethane with bromine

This reaction is called radical substitution. It takes place in three stages:

1 **Initiation** – homolytic bond fission of bromine to form bromine radicals, using UV light or heat.

2 **Propagation** – repeated steps where one radical reacts and another radical forms.

3 **Termination** – combination of any two radicals to form a stable product.

The equations for each step

Step	Equations
initiation	$Br_2 \rightarrow 2Br^\bullet$
propagation	$C_2H_6 + Br^\bullet \rightarrow C_2H_5 + HBr$ $^\bullet C_2H_5 + Br_2 \rightarrow C_2H_5Br + Br^\bullet$
termination	$2Br^\bullet \rightarrow Br_2$ $2^\bullet C_2H_5 \rightarrow C_4H_{10}$ $^\bullet C_2H_5 + Br^\bullet \rightarrow C_2H_5Br$

The termination steps produce a number of products, including reforming the halogen. Further substitutions can occur, forming other haloalkanes such as 1,1-dibromoethane, $CHBr_2CH_3$. Hence this reaction is of limited use in synthesis, as it's difficult to control in order to produce the desired product.

Chlorine will react with ethane in the same way as bromine.

Worked example

Write equations to show how 1,2-dibromoethane can form during the radical reaction of bromine with ethane.

$CH_2BrCH_3 + Br^\bullet \rightarrow CH_2BrCH_2^\bullet + HBr$

$CH_2BrCH_2^\bullet + Br^\bullet \rightarrow CH_2BrCH_2Br$

Write an equation for the combustion of butane, forming only toxic gas and water.

$C_4H_{10} + 4\tfrac{1}{2}O_2 \rightarrow 4CO + 5H_2O$

Bromoethane, CH_2BrCH_3 formed in the reaction may collide with the bromine radical, forming a new radical in a **propagation** step.

The new radical may then collide with another bromine radical in a **termination** step forming 1,2-dibromoethane.

 The question gives a clue to highlight that this is **incomplete** combustion by referring to 'toxic gas'. This indicates that the product containing carbon must be carbon monoxide, CO.

Now try this

1 Write equations for the mechanism of the reaction of chlorine with methane in the presence of UV light. **(4 marks)**

2 Write an equation for the combustion of propane, forming a black solid and water only. **(2 marks)**

Bonding in alkenes

Alkenes are a family of unsaturated hydrocarbons with the general formula C_nH_{2n}. They all contain at least one carbon–carbon double bond.

Ethene

Ethene is the simplest member of the alkene family. Its displayed formula is shown below.

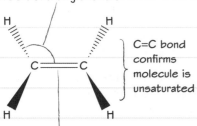

HCC bond angle is 120°

C=C bond confirms molecule is unsaturated

Remember alkenes are a *homologous series* – compounds with the same functional group that differ only by a CH_2 group.

Double bond consists of a **sigma** bond and a **pi** bond

General formulae

A general formula is a way of representing, with symbols, the composition of any member of an entire homologous series.

For instance, the general formula for alkenes, C_nH_{2n} means you can deduce the molecular formula of any member of the alkenes if you know how many carbon atoms it has.

So an alkene with 4 carbons will have the molecular formula C_4H_8.

Formation of a sigma (σ) bond

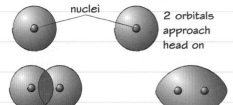

nuclei

2 orbitals approach head on

Orbitals overlap to form sigma bond

Increased electron density between nuclei

Formation of a pi (π) bond

2 p-orbitals approach sideways on

p-orbitals overlap to form pi bond

Increased electron density above and below plane of molecule

Worked example

Describe what is meant by the term 'pi (π) bond'.
(2 marks)

A pi bond is a region of space occupied by a bond pair of electrons, above and below the plane of the molecule, formed by sideways overlap of two p orbitals.

You could draw labelled diagrams like the ones above to show what is meant by pi bond.

Worked example

Is a pi bond stronger or weaker than a sigma bond? Explain your answer.
(2 marks)

Weaker, as the regions occupied by the electrons are further from the nuclei so the electrostatic attraction to the nuclei is weaker.

Take care to emphasise that the attraction is electrostatic. Describing other forces, for example magnetic, is incorrect and could lose you a mark.

Now try this

Explain the shape around each carbon atom in ethene. **(2 marks)**

The answer must state the shape then explain it using **electron pair repulsion** theory.

Remember the double bond counts as a single area of **electron density**.

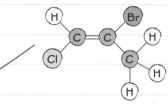
Stereoisomerism in alkenes

Stereoisomers are compounds with the same structural formula but their atoms have a different arrangement in space.

Spatial arrangements in alkenes

Alkenes can show stereoisomerism if they have two different groups attached at each end of the C=C bond.

The haloalkene, $CHClCBrCH_3$, has two possible spatial arrangements.

Arrangement 1

These two groups of atoms show the different spatial arrangements. The double bond has restricted rotation, so the groups are locked in position. Each different arrangement is an alternative stereoisomer.

Arrangement 2

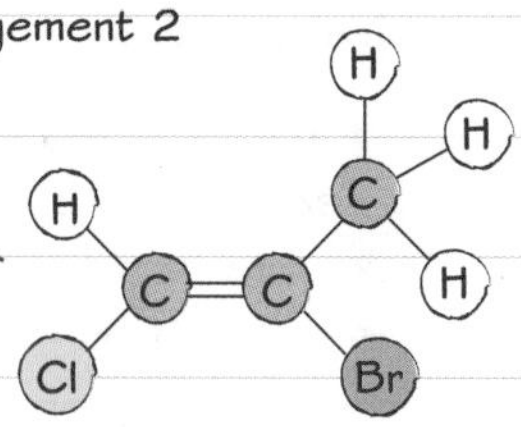

Naming stereoisomers

You can name stereoisomers using E/Z nomenclature by applying the Cahn–Ingold–Prelog rules.

The four groups attached to the C=C bond are prioritised according to the sum of their atomic masses. The highest mass has the highest priority.

E-2-bromo-1-chloropropene

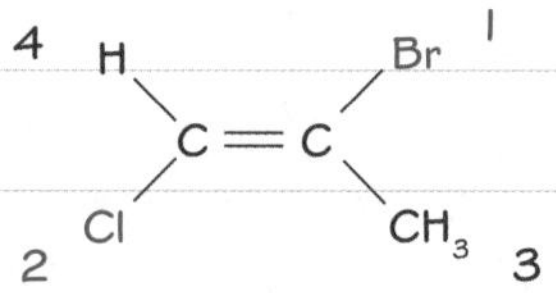

$$4 \quad H \qquad Br \quad 1$$
$$C=C$$
$$2 \quad Cl \qquad CH_3 \quad 3$$

The two highest priority groups are opposite. This is the *E*-isomer. The German word for opposite is *entgegan*.

Z-2-bromo-1-chloropropene

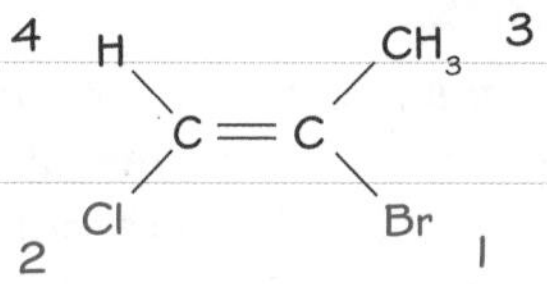

$$4 \quad H \qquad CH_3 \quad 3$$
$$C=C$$
$$2 \quad Cl \qquad Br \quad 1$$

The two highest priority groups are together, on the same side. This is the *Z*-isomer. The German word for together is *zusammen*.

Worked example

Name this isomer using the Cahn–Ingold–Prelog rules:

$$H \qquad\qquad CH_3$$
$$C=C$$
$$H_3C \qquad\qquad H$$

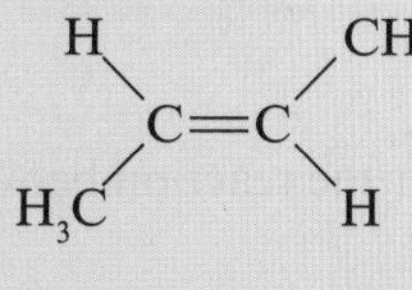

(2 marks)

E-but-2-ene

In this case the two highest priority groups, the two CH_3 groups, are opposite so this is the *E*-isomer.

As two of the groups attached to each carbon are identical, the *cis–trans* naming system can be used here.

Hence this isomer is also known as *trans* but-2-ene as the two identical groups are opposite.

If the 2 identical groups on each carbon are on the same side it would be called *cis* but-2-ene

Now try this

1 Draw the *cis* isomer of but-2-ene. **(1 mark)**

2 Use the Cahn–Ingold–Prelog rules to name this isomer.

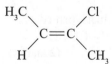

$$H_3C \qquad\qquad Cl$$
$$C=C$$
$$H \qquad\qquad CH_3$$

(1 mark)

3 Can the *cis–trans* system be used to name the isomer in question 2? Explain your answer.

(2 marks)

Addition reactions of alkenes

Addition reactions occur when two reactants combine together to form one product.

Key facts about the addition reactions of alkenes

☑ The **electron-rich** C=C bond in alkenes is susceptible to attack by **electrophiles**.

☑ Electrophiles are species that can accept a pair of electrons.

☑ Electrophiles are either positive ions or polar molecules.

☑ Some species can become an electrophile by passing near to a C=C double bond, as its high electron density is enough to induce a **dipole**.

Key reactions

You are expected to know the products formed and reactants needed for each of these reactions.

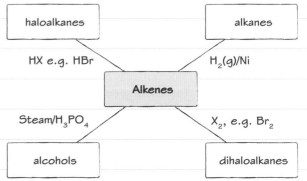

haloalkanes		alkanes

HX e.g. HBr   H₂(g)/Ni

Alkenes

Steam/H₃PO₄   X₂, e.g. Br₂

alcohols		dihaloalkanes

Practise drawing and naming the products for a range of different alkenes.

Addition reaction mechanisms

This is the **mechanism** for an electrophilic addition reaction – a series of steps showing how the reactants combine to form the products. The bond breaking process in this case is heterolytic fission.

Remember that a **curly arrow** shows the direction in which electron pairs move. Although the electrophile or alkene may differ, the basic steps of the mechanism remain the same, so don't panic if you get asked to show the mechanism with different reagents.

Electron pair from double bond moves to Br$^{\delta+}$

Intermediate carbocation forms

Product forms

Br$^{\delta+}$

Br$^{\delta-}$

Bond pair moves onto Br$^{\delta+}$ to form bromide ion

:Br⁻

Lone pair moves from bromide ion to carbocation, forming a bond pair

Worked example

(a) Name the two possible products formed when HBr(g) reacts with CH₂=CHCH₃. **(2 marks)**

1-bromopropane and 2-bromopropane

(b) Explain which of your two answers is likely to be the major product. **(3 marks)**

The major product is likely to be 2-bromopropane as the intermediate formed in the mechanism is a secondary carbocation. This is because the methyl group attached to the positive carbon pushes the electron bond pair towards this carbon.

Hence the secondary carbocation is more stable than the primary carbocation formed when 1-bromopropane is the product.

Now try this

A student proposed this mechanism for the reaction between bromine and but-1-ene.
Suggest two errors with this proposal. **(2 marks)**

H$^{\delta+}$— Br$^{\delta-}$

:Br⁻

It's always a good idea to make comparative statements in this kind of answer, e.g. 'more stable'. This shows you have considered both possibilities and made a judgement.

Formation and disposal of polymers

Polymers are useful materials manufactured from simple molecules called **monomers**, that are often derived from oil. Alkenes form polymers by addition reactions and chemists play many roles in providing sustainable ways to process used polymers.

Formation of polymers using addition reactions

Addition polymers form when the pi bond in a monomer breaks. This allows these electrons to form a bond with a neighbouring carbon. This can occur many times, forming very long chains. For example ethene will form (poly)ethene.

Note the open bonds at the end, showing that the chain continues.

It's not possible to draw the whole polymer as it is huge, so a simplified version of the formula is used, for example:

$$n \quad \underset{Y}{\overset{W}{\diagdown}} C = C \underset{Z}{\overset{X}{\diagup}} \longrightarrow \left[\begin{array}{cc} W & X \\ | & | \\ -C-C- \\ | & | \\ Y & Z \end{array} \right]_n$$

The letter 'n' represents a large number. W, X, Y and Z can be any atom or group of atoms.

How we can use polymers more sustainably

Polymers generally are derived from crude oil so their continued use will deplete a finite resource. Chemists play a key role in using this resource in a manageable way, while developing alternatives that do not rely on oil.

Incineration - produces useful energy but requires careful treatment of potentially harmful emissions e.g. HCl produced when poly(chloroethene) burns.

Biodegradable polymers - chemists are developing polymers from materials such as corn starch. These have the advantage of being biodegradable as well as using sustainable raw materials.

Reuse - plastic containers can often be rinsed and used again.

Recycle - requires sorting, cleaning and melting down for re-shaping. Often the 'new' plastic is weaker than the original material.

Photodegradable polymers - exposure to UV light breaks bonds in polymers, forming smaller molecules.

Producing useful starting materials - breaking down plastics into smaller molecules in a process similar to cracking produces small molecules that can be used as starting materials in the petrochemical industry. Such molecules are called **feedstock**.

Many students struggle when presented with the structural formula of the monomer, especially if it contains more than 2 carbon atoms. A useful tip is to display the monomer first, in this format.

$$\underset{Y}{\overset{W}{\diagdown}} C = C \underset{Z}{\overset{X}{\diagup}}$$

This will help you visualise which groups **will not** change in the final polymer.

Worked example

Draw the formula for the polymers formed from these monomers.

(a) CH_2CHOH
(1 mark)

(b) CH_3CHCH_2
(1 mark)

Now try this

Describe two disadvantages and two advantages of the use of addition polymers and explain how chemists may help make their use more sustainable. **(6 marks)**

Exam skills 5

This exam-style question uses knowledge and skills you have already revised. Have a look at page 65 for a reminder about **polymers** and pages 21 and 22 for a reminder about shapes of molecules.

Worked example

The monomer 1,1,2,3,3,3-hexafluoropropene can be used to make a polymer used to prevent ice from building up on solid surfaces such as an aeroplane's wings.

F   F
| |
F—C—C=C
| | |
F   F   F

(a) Write an equation to show the formation of the addition polymer formed from this monomer.

(3 marks)

$$n \quad \begin{matrix} F & & F \\ | & & | \\ F—C—C=C— \\ | & | & | \\ F & F & F \end{matrix} \longrightarrow \begin{bmatrix} CF_3 & F \\ | & | \\ —C—C— \\ | & | \\ F & F \end{bmatrix}_n$$

(b) State the bond angle around each carbon atom in the polymer above. Explain your answer.

(2 marks)

109.5° as each carbon atom has four bonds. The four bond pairs of electrons repel each other equally given the angle associated with a perfect tetrahedron.

(c) Polymers often use finite raw materials in their manufacture, such as crude oil. Explain how society, with the help of chemists, can ensure polymer waste is managed sustainably. **(4 marks)**

Society could recycle more waste plastics, so less crude oil is used to manufacture new plastics or less non-biodegradable material goes into landfill.

Chemists could help improve incineration techniques to minimise production of toxic products **and** recover energy from combustion.

Chemists could process waste plastics to use as feedstock (raw material) in the chemical industry.

Chemists are developing biodegradable polymers that use plant-based raw materials and do not persist in the environment.

You'll find it helpful when drawing the structure of polymers to redraw the monomer in a simplified format as shown. This will help you to draw the repeat unit.

Notice the polymer must have continuation bonds extending out from the repeat unit. Only one repeat unit, multiplied by n, is shown in the equation. If more than one repeat unit is shown inside the square brackets then this equation would not balance.

An equation must balance, hence the use of 'n' on the left to represent the number of monomers, which varies depending on the polymer in question. It appears again on the right to show that the polymer consists of that many monomers.

'State' here simply means to recall a piece of information. Although you may not be familiar with this specific polymer you will have studied the shapes of molecules with similar bonding so you can use that knowledge.

'Explain' requires you to exemplify a point. Here, reasoning based on electron pairs repelling each other is used to back up the stated bond angle.

When predicting bond angles and shapes you may find it helpful to draw a dot-and-cross diagram to help you determine the number of bond pairs and lone pairs around an atom.

The use of 'explain' means you have to do a bit more than just describe sustainable management of polymer waste – you must say **why** your answers result in sustainable management, linking your reasons clearly to your answers.

Simple generic responses, such as 'recycling causes less pollution' are unlikely to gain credit.

The properties of alcohols

The structure of alcohols helps explain some of their properties, such as volatility and solubility in water, compared to alkanes.

Hydrogen bonding in ethanol

An oxygen atom in water could also interact with the electropositive hydrogen atom in the hydroxyl group. This explains why ethanol is soluble in water.

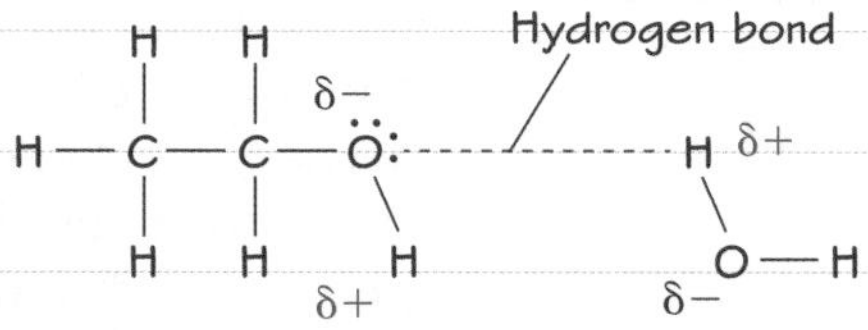

An electropositive hydrogen, for instance in water, can form a **hydrogen bond** to the oxygen atom in ethanol.

The greater electronegativity of the oxygen atom in the hydroxyl group allows it to pull the OH bond pair towards itself, making the O-H bond polar.

Classifying alcohols

Alcohols are classified as primary, secondary or tertiary, depending on their structure.

This is a **primary** alcohol. The carbon attached to the hydroxyl group is attached to only one other carbon atom.

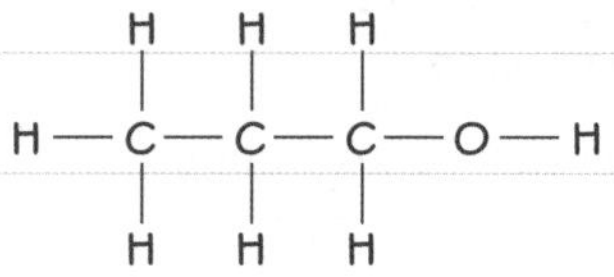

Propan-1-ol

This is a **secondary** alcohol. The carbon attached to the hydroxyl group is attached to two other carbon atoms.

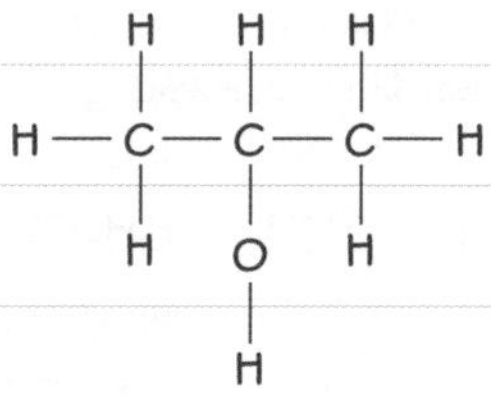

Propan-2-ol

This is a **tertiary** alcohol. The carbon attached to the hydroxyl group is attached to three other carbon atoms.

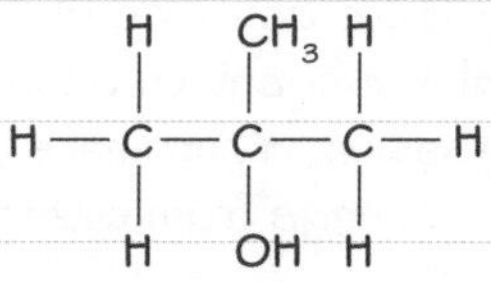

Methylpropan-2-ol

Be careful! The numbering used in the names of alcohols refers to the position of the –OH group, not the classification. Hence methylpropan-2-ol is NOT secondary, as the '2' refers to the –OH being positioned on carbon number 2.

Worked example

Compare the solubility of pentan-1-ol in water to that of ethanol. Explain your answer. **(3 marks)**

- Both pentan-1-ol and ethanol can form hydrogen bonds to water.
- Pentan-1-ol will be less soluble than ethanol due to its longer hydrocarbon chain.
- The hydrocarbon chain cannot interact with water as the C-H bonds are non-polar.

Now try this

A bank note soaked in a 50:50 mixture of ethanol and water easily ignites. Suggest why the note is left undamaged once all the ethanol has burnt. **(2 marks)**

Combustion and oxidation of alcohols

Alcohols are often flammable and undergo combustion readily in air. Some alcohols can also be oxidised using certain oxidising agents and conditions.

Combustion of alcohols

Alcohols will burn in air to form carbon dioxide and water. The general equation for the reaction is

$$C_nH_{2n+1}OH + (1.5n)O_2 \rightarrow (n + 1)H_2O + nCO_2$$

so for example:

$$C_2H_5OH + 3O_2 \rightarrow 3H_2O + 2CO_2$$

Although all alcohols can burn in oxygen, only primary and secondary alcohols can be oxidised by oxidising agents. Make sure you can apply your knowledge of covalent bonding to explain why tertiary alcohols will burn but cannot be readily oxidised by oxidising agents such as acidified potassium dichromate(VI).

Practical skills ## Oxidation of alcohols

Water out

Primary alcohols can be partially oxidised to form aldehydes.

Reaction mixture

Aldehyde

Heat

A simple distillation is used to remove the product as soon as it forms to prevent further oxidation.

The reaction mixture consists of the primary alcohol and an oxidising agent, which is often acidified potassium dichromate(VI), $K_2Cr_2O_7 / H^+$.

The equation for such reactions can be simplified using [O] to represent oxygen from the oxidising agent. For example:

$$CH_3CH_2OH + [O] \rightarrow CH_3CHO + H_2O$$

Primary alcohols can be completely oxidised to form carboxylic acids, and secondary alcohols to form ketones.

Condenser

Water in

Heating under reflux is used so the reaction mixture can be heated for longer periods so the reaction goes to completion. The condenser makes sure any vapour condenses back into the flask to minimise the release of harmful or flammable substances.

Reaction mixture

Anti-bumping granules

Heat

$$CH_3CH_2CH_2OH + 2[O] \rightarrow CH_3CH_2COOH + H_2O$$
propan-1-ol   propanoic acid

$$CH_2CH(OH)CH_3 + [O] \rightarrow CH_3COCH_3 + H_2O$$
propan-2-ol   propanone

The reaction mixture consists of an alcohol and an **excess** of oxidising agent to ensure complete oxidation. If acidified potassium dichromate (VI) is used, there will be a colour change from orange to green.

There may be more unusual examples in exams.

Just identify any primary or secondary alcohol groups and they become carboxylic acids and ketones respectively. You can ignore any tertiary alcohol groups as they do not readily oxidise.

Worked example

Draw the structure of the product formed when the alcohol shown is heated under reflux with excess acidified potassium dichromate(VI).

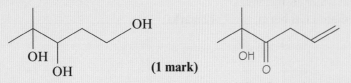

(1 mark)

Now try this

Write the equation for the incomplete combustion of ethanol, forming a toxic gas and water only.

(2 marks)

More reactions of alcohols

Alcohols are useful in organic chemistry as they can be converted into a number of other compounds. You need to be able to recall and apply the reactions shown here and recognise whether the C–O bond or the O–H bond is broken in a reaction.

Elimination of water from alcohols

Either of the hydrogen atoms adjacent to the functional group carbon can be removed.

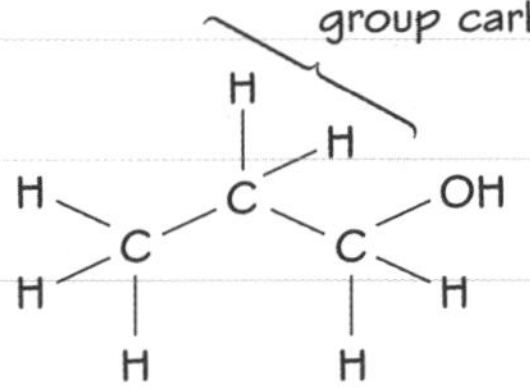

propan-1-ol

H_3PO_4 catalyst
heat

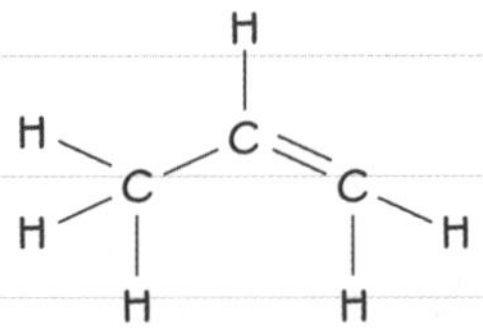

propene

Sulfuric acid, H_2SO_4, can also be used as a catalyst.

Substitution of hydroxyl group by halide ions

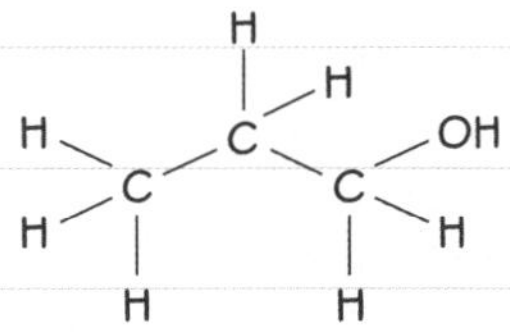

propan-1-ol

NaX
H_2SO_4

NaX could be any sodium halide

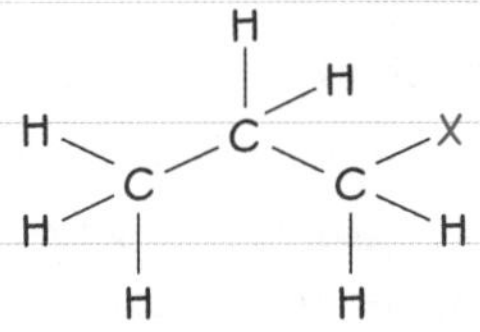

halopropane

Worked example

Name the organic products of the reaction when butan-2-ol is heated in the presence of a phosphoric acid catalyst. **(3 marks)**

but-1-ene

E-but-2-ene

Z-but-2-ene

Remember that one of the hydrogen atoms removed comes from the carbon next to the functional group carbon. In the case of butan-2-ol, this could mean a hydrogen is removed from the first carbon (forming but-1-ene) or the third carbon (forming either of the stereoisomers of but-2-ene).

Now try this

1 Name the organic product formed when pentan-1-ol reacts with potassium chloride, in the presence of an acid catalyst. **(1 mark)**

2 State the conditions required to form an alkene from an alcohol. **(2 marks)**

3 Write equations for the
  (a) partial oxidation of methanol
  (b) complete oxidation of butan-2-ol
  (c) complete oxidation of ethan-1,2-diol. **(3 marks)**

Nucleophilic substitution reactions of haloalkanes

Haloalkanes are organic compounds where one or more of the hydrogens in an alkane are substituted by a halogen. You will have come across them earlier as products from radical substitution reactions of alkanes and electrophilic addition reactions of alkenes.

Mechanism for hydrolysis of a haloalkane

A reaction of haloalkanes is the substitution of the halogen atom by a hydroxide ion, either from an alkali such as sodium hydroxide, or from water. This forms an alcohol and a halide ion. While the reaction is often referred to as a hydrolysis reaction, i.e. a reaction with water, its mechanism is an example of a **nucleophilic substitution** reaction. A *nucleophile* is defined as a species that can donate an electron pair. In this case the nucleophile is the hydroxide ion.

The C–I bond breaks and the bond pair moves onto the iodine, forming an iodide ion. The lone pair on the hydroxide ion allows it to form a covalent bond to the electron-deficient carbon.

There are some key points to keep in mind when drawing a reaction mechanism:

- Always show the partial charges on relevant polar bonds. In this case it's the C–I bond.

- Make sure any lone pairs are clearly shown on the correct atom in a nucleophile.

- Remember that an arrow denotes movement of a pair of electrons, so will always point away from a nucleophile, as in this example, but towards an electrophile.

🧪 Practical skills — Rate of hydrolysis of primary haloalkanes

A simple experiment can be carried out to find the rate of hydrolysis of different haloalkanes by warming three test tubes each containing a different haloalkane with aqueous silver nitrate solution. The water in the solution provides the hydroxide ions and as soon as halide ions are released a silver halide precipitate forms.

Haloalkane	Colour of ppt	Time to form ppt / s
1-chloropropane	white	150
1-bromopropane	cream	130
1-iodopropane	yellow	90

The solvent used for this reaction is often a mixture of ethanol and water. The water acts as the nucleophile but many haloalkanes do not dissolve in water alone. Ethanol is added to help ensure the haloalkane dissolves and hence reacts.

You can see that the order of the rate of hydrolysis from fastest to slowest is:

iodoalkane > bromoalkane > chloroalkane

The reason for this is the relative strengths of the C–X bond. The C–I bond is the weakest, so breaks most easily. Hence the yellow precipitate of silver iodide is seen first.

Be careful! A common mistake is to try to justify this experiment in terms of electronegativity of the halogen but experimental data does not fit that explanation. If electronegativity was the key factor in this experiment the chloroalkane would react the fastest. However, the experiment tells us the iodoalkane reacts fastest. Hence **bond enthalpy** must be the key factor.

Worked example

Describe what variables you would need to control to ensure an experiment to compare the rates of hydrolysis of haloalkanes gives valid results.   **(3 marks)**

You would need to control the length of carbon chain in the haloalkane, the volume/concentration of silver nitrate solution used and the amount of haloalkane used.

You could also have mentioned the temperature (of water bath used for heating) and classification of haloalkane (primary etc.) in your answer.

Now try this

Write the mechanism for the hydrolysis of 1-bromobutane, using skeletal formulae.   **(3 marks)**

Preparing a liquid haloalkane

Practical skills   Preparing organic compounds such as haloalkanes and alcohols requires a number of practical skills as well as recall of the reagents and conditions required. Such skills include knowing how to set up apparatus to allow reagents to be heated safely and how to separate a liquid product from the rest of a reaction mixture.

Using Quickfit apparatus to heat under reflux

Some iodoalkanes can be prepared by heating an alcohol with appropriate reactants under reflux.

The heating technique is similar to that used when completely oxidising alcohols (see page 68).

Heating under reflux is a method of gently heating a reaction mixture for a long time. The use of the condenser ensures reactants and products do not escape as they are needed for the reaction and may be harmful or flammable.

The glassware used is precisely made so that a seal forms when pieces are connected. This prevents escape of any vapour.

Water out

Water in

The water flowing inside the condenser jacket cools down any vapour, condensing it back to a liquid.

The condenser works more effectively if the water feeds into the jacket at the bottom. This prevents air bubbles from forming in the jacket.

Reaction mixture

Electric heater

An electrical heater or water bath is often used to avoid a naked flame near any flammable reactants or solvents.

Extracting and purifying the liquid product

The organic product has to be separated from the reaction mixture and purified. This involves washing to remove impurities, drying and finally purifying by distillation.

Transfer reaction mixture to separating funnel and add washing agent	Shake funnel gently, releasing pressure as necesary	Remove unwanted layer. Transfer organic layer to flask, add drying agent	Filter off drying agent, then distil off the product

The washing agent is normally water, but sometimes a dilute acid or base is used, for instance to neutralise any catalyst.

Suitable drying agents include anhydrous calcium chloride or anhydrous magnesium sulfate.

Check boiling point to confirm purity.

Worked example

Draw the apparatus used to distil off a liquid product.   **(3 marks)**

A simple cross-section diagram of the apparatus is all that is required, not a three-dimensional masterpiece! Common mistakes include students sealing their entire apparatus, leaving **large** gaps where glassware joins and having the water flowing in the wrong direction.

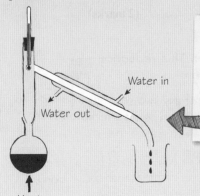

Water in

Water out

Heat

Key points:
- flask, heat, still head (connects flask to condenser)
- correctly placed thermometer
- condenser with correct water flow and collection vessel.

Now try this

Explain how the washing, drying and distillation stages help ensure a pure product is obtained, when preparing an organic liquid.

(3 marks)

Haloalkanes in the environment

Haloalkanes have been used for a variety of purposes, including aerosol propellants and refrigerants. Their use has lessened as chemists explained their impact on the environment and looked for alternatives.

Chlorofluorocarbons – CFCs

The chlorofluorocarbon CF_2Cl_2, commonly known as Freon-12, has the structure

dichlorodifluoromethane

The strength of the C–F and C–Cl bonds means that such CFCs are chemically inert. This property means they are stable and non-toxic.

The ozone layer

Ozone is a form of oxygen with the formula O_3. It is toxic and in the lower parts of the atmosphere it can contribute to the formation of photochemical smog. However, high in the atmosphere it plays a vital role in absorbing UV radiation, reducing the levels that reach the Earth's surface. This is important as exposure to high levels of UV can lead to skin cancer and cataracts.

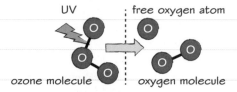

UV · free oxygen atom

ozone molecule · oxygen molecule

The ozone problem

As CFCs are inert, once released into the air they could diffuse into the upper atmosphere unchanged. There they are exposed to increased levels of UV light. This energy causes **homolytic bond fission** in the haloalkanes, forming free radicals.

$$CF_2Cl_2 \rightarrow CF_2Cl^{\bullet} + Cl^{\bullet}$$

The free chlorine radical can then react with ozone, converting it to oxygen, which is unable to absorb UV radiation.

$$Cl^{\bullet} + O_3 \rightarrow {}^{\bullet}ClO + O_2$$

This reaction forms another free radical $^{\bullet}ClO$, which also reacts with ozone, regenerating the $Cl^{\bullet}$ free radical.

$${}^{\bullet}ClO + O_3 \rightarrow Cl^{\bullet} + 2O_2$$

Propagation

This ongoing propagation of free radicals means that the production of a single chlorine free radical can be responsible for converting thousands of ozone molecules into oxygen.

Other radicals and ozone

Nitrogen oxides, produced in the extreme high temperatures of car and aircraft engines, also form free radicals that can react with ozone. The equations below summarise such a series of reactions.

Initiation step: $NO_2 \rightarrow NO^{\bullet} + O^{\bullet}$

Propagation step 1: $NO^{\bullet} + O_3 \rightarrow {}^{\bullet}NO_2 + O_2$

Propagation step 2: $NO_2{}^{\bullet} + O_3 \rightarrow NO^{\bullet} + 2O_2$

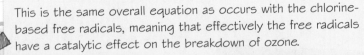

This is the same overall equation as occurs with the chlorine-based free radicals, meaning that effectively the free radicals have a catalytic effect on the breakdown of ozone.

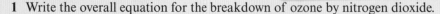

Worked example

1 Write the overall equation for the breakdown of ozone by nitrogen dioxide. **(1 mark)**

$$2O_3 \rightarrow 3O_2$$

2 Suggest an equation for a reaction that will prevent further depletion of ozone by free radicals. Explain your choice of reaction. **(2 marks)**

$$2{}^{\bullet}ClO \rightarrow Cl_2 + O_2$$

This forms species without an unpaired electron, which are less reactive. This reduction in the number of free radicals slows down the overall depletion of ozone.

You could choose the combination of any two relevant free radicals. This combination of free radicals is known as a **termination** step.

Now try this

Using your knowledge of bond strengths suggest an alternative to CF_2Cl_2 as a refrigerant which will have less impact on the ozone layer. **(2 marks)**

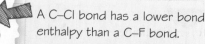

A C–Cl bond has a lower bond enthalpy than a C–F bond.

Organic synthesis

Knowledge of the reactions of different functional groups can be used to put together reaction sequences to convert a starting material to a product in a series of steps. Products can then be tested to identify them.

Tackling problems

To tackle synthetic problems it's vital that you have a thorough knowledge of all the organic reactions you have studied in the course.

For each reaction make sure you can recall:

 reagents

 conditions

 reaction type

 details of mechanism where required.

In Year 1 you will be expected to apply this knowledge to propose simple synthetic routes, consisting of two steps.

Practical skills — Synthesis of cyclohexanol from cyclohexene in two steps

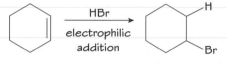

cyclohexene   bromocyclohexane   cyclohexanol

The second reaction is referred to as a nucleophilic substitution but it is also known as a hydrolysis. It's important that the NaOH is dissolved in water, otherwise side reactions would form other unwanted products.

Although the specification only requires two-step processes in Module 4, it requires you to consider up to three-step processes with up to three steps for the full A level. This will require knowledge of all organic reactions across the entire course.

Worked example

Outline a synthesis to produce butanoic acid from 1-chlorobutane in two steps. Include details of reagents, conditions and reaction types. Describe observations you might see or tests you could carry out at the end of each step to confirm you have made the desired product.   **(8 marks)**

1-chlorobutane → NaOH (aq) nucleophilic substitution → butan-1-ol → oxidation, excess acidified potassium dichromate ✓, heat under reflux ✓ → butanoic acid

Test for butan-1-ol – oxidising agent, acidified potassium dichromate, should turn from orange to green.

Test for butanoic acid – add a small amount of sodium carbonate solution, should see effervescence or fizzing.

A useful approach to this sort of question is to break it down in smaller chunks before you start. Highlighting or numbering will help.

Now try this

Describe how to test for the following functional groups, including the expected positive result.

(a) Alkene   **(2 marks)**

(b) Aldehyde   **(2 marks)**

(c) Haloalkane   **(2 marks)**

When describing colour changes it is good practice to make it obvious what the colour was both before and after any reactions – and remember if something looks like water, it's **colourless, not clear.**

Exam skills 6

This exam-style question uses knowledge and skills you have already revised. Look at pages 55, 70 and 71 for a reminder of key terms in organic chemistry, the reactions of haloalkanes and the preparation and purification of organic liquids.

Worked example

Bromoethane, CH_3CH_2Br, can be made by slowly adding concentrated sulfuric acid to a round-bottomed flask containing ethanol. The flask is kept cool in an ice-bath. Solid potassium bromide is then added. The impure product is then removed by a simple distillation.

(a) Suggest two impurities, not including water, found in the reaction mixture and explain how they formed.

(4 marks)

Possible impurities include:

- ethanol, as the reaction has not gone to completion
- hydrogen bromide formed as a by-product in the reaction
- bromine, as sulfuric acid oxidises bromide ions
- sulfur dioxide, formed when the sulfuric acid oxidises bromide ions.

(b) Many of the impurities are water soluble. Describe how to remove them, leaving a dry sample of the organic liquid formed.

(3 marks)

Add the organic liquid to a separating funnel. Add distilled water, stopper and shake gently. Release any pressure formed by opening the tap slightly from time to time. Separate the layers, collecting each one in a different conical flask.

Add a drying agent such as anhydrous magnesium sulfate to the organic layer to absorb any remaining water. Leave for a few minutes, then filter off the anhydrous magnesium sulfate.

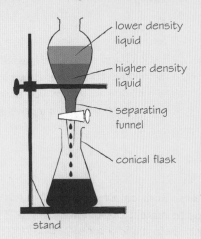

lower density liquid

higher density liquid

separating funnel

conical flask

stand

This question needs you to look beyond the organic reaction and use your knowledge of oxidation and reduction to suggest possible impurities. As you get further into the course there will be more of an expectation that you link together concepts from different units.

Practical skills Knowledge of certain key practical skills is specifically required by the specification. The techniques used to prepare and purify an organic liquid are one such example.

Practical skills You should check the densities of the liquids to determine which layer is the organic product and which is the water-based layer.

Practical skills You need to be able to describe how to set up apparatus to heat under reflux. You may find it useful to practise drawing such apparatus and spotting the common errors students make, such as leaving gaps between joined pieces of glassware.

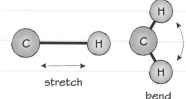
AS & A level
Module 4

Infrared spectroscopy

Infrared spectroscopy helps chemists to identify compounds and monitor products of reactions by observing how molecules interact with infrared light.

Infrared radiation

Molecules will absorb infrared radiation as the frequency range of infrared closely matches the frequency of vibration of covalent bonds.

Different bonds vibrate at slightly different frequencies, allowing chemists to determine which bonds are present in a sample.

stretch

bend

Some gases in the atmosphere absorb infrared radiation and so prevent some of this energy escaping out into space. This is called the Greenhouse Effect. These gases include carbon dioxide, water vapour and methane. Release of compounds such as CO_2, from burning of fossil fuels, has led many scientists to link the increase in such gases to the increase in average global temperatures, known as global warming.

Infrared spectra

If a compound being studied has had its spectrum recorded previously it can be identified by comparing the spectrum taken to a database. The absorptions on the right-hand side of the spectrum, called the 'fingerprint' region, are especially useful for this.

What you need to know

You are not expected to know precisely how an infrared spectrometer works. However, you may be given an infrared spectrum and be expected to use it as part of a series of techniques to identify a compound, especially with reference to the absorptions due to O–H and C=O bonds. You don't have to memorise all wavenumbers relevant to particular bonds and groups, you will be given this information on the Data Sheet.

An absorption at around $3000\,cm^{-1}$ is evident on all spectra as it's caused by the vibration of C–H bonds – so no help in identifying the molecule!

The broad absorption at 3000–$3500\,cm^{-1}$ shows the presence of an O–H bond. The trough is wider than normal due to hydrogen bonding between the molecules.

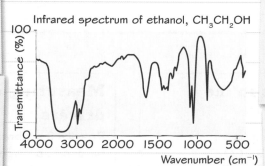
Infrared spectrum of ethanol, CH_3CH_2OH

The stretches due to the C–O bond are used by forensic scientists looking for the presence of ethanol on the breath of a suspected drink driver.

The units of the 'wavenumbers' on the x-axis of the spectrum are cm^{-1} which is proportional to the energy of the vibration.

Worked example

The IR spectrum of a compound with the molecular formula $C_3H_6O_2$ is shown below. Identify the compound, explaining how you arrived at your answer.   **(3 marks)**

Broad peak in region 2500–$3000\,cm^{-1}$ suggests an –OH group. Broader than that expected for an alcohol, but matches expected absorption for the –OH group in a carboxylic acid.

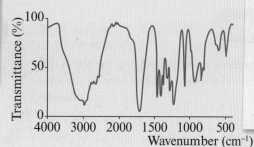

Absorption peak at around $1700\,cm^{-1}$ suggests a carbonyl group but can't be an aldehyde or ketone due to presence of –OH absorption.

Contains carboxylic acid –OH group and a carbonyl group. Molecular formula shows three carbon atoms so compound is $CH_3CH_2CO_2H$.

Now try this

Use your OCR Data Sheet to identify the key absorptions in the IR spectrum of 2-hydroxybutanal and explain how you could use the spectrum to differentiate between this compound and butanoic acid.   **(3 marks)**

Uses of infrared spectroscopy

Infrared spectrometers have a wide variety of uses both in and outside the laboratory.

Identifying unknown substances

While it's unlikely that an infrared spectrum on its own could be used to positively identify a compound, it is often used specifically to help identify particular functional groups in organic chemistry. For instance look at this IR spectrum of ethyl ethanoate.

Lack of an –OH stretch rules out a carboxylic acid. Beware of the C–H stretch. This is in the same region as some O–H stretches but, unlike the O-H stretch, it is not broad.

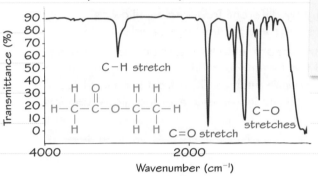

The presence of a C–O stretch at about 1000 cm⁻¹ rules out the possibility of an aldehyde or ketone.

The C=O stretch at about 1750 cm⁻¹ could be an ester, carboxylic acid, aldehyde or ketone.

Characteristic infrared absorptions in organic molecules

Bond	Location	Wavenumber / cm⁻¹
C–O	alcohols, esters, carboxylic acids	1000–1300
C=O	aldehydes, ketones, carboxylic acids, esters, amides	1640–1750
C–H	organic compound with a C–H bond	2850–3100
O–H	carboxylic acids	2500–3300 (very broad)
N–H	amines, amides	3200–3500
O–H	alcohols, phenols	3200–3550 (broad)

Monitoring gases in the atmosphere

As well as being used to monitor atmospheric gases such as CO_2, concentrations of gases in vehicle exhaust emissions can be measured using IR.

Such units can measure precise amounts of gases such as carbon monoxide and nitrogen oxides using IR absorptions.

Measuring ethanol in the breath of drivers

Results from breathalysers are accurate enough to be used as evidence, reducing the need to take blood samples from suspected drunk drivers. Many machines are now small enough to be used at the roadside. As well as using IR to identify key bonds in ethanol such as the C–O bond, they often use fuel cells in conjunction, to check ethanol levels.

Worked example

Suggest three advantages of using IR technology in breath testing compared to blood testing, when measuring the levels of ethanol in drivers. **(3 marks)**

The police will no longer need a medical practitioner present to collect evidence suitable for use in court. It also means evidence is available quickly without having to wait for results of blood tests.

The IR breath test is non-invasive, which is less stressful for any suspects who are afraid of needles or blood, and hence easier for the police to manage.

Now try this

Suggest three problems which atmospheric gases may cause that chemists may want to monitor. **(3 marks)**

These developments have only come about because the technology now gives consistently accurate and reliable data quickly and can be used by trained members of the Police Service.

Mass spectrometry

Mass spectrometry can be used to help identify the structure of a molecule.

The mass spectrum

Mass spectrometers form ions from any molecule placed inside them, and record the mass of those ions. The data from a mass spectrum can be used as part of the evidence to help determine the structure of a compound.

The overall pattern in the mass spectrum is like a fingerprint so can be used to match against spectra that have already been recorded.

The peak with the highest m/z value is due to the removal of one electron from the molecule. The ion formed is called the molecular ion. The value of m/z is equivalent to the relative molecular mass of the compound. For pentane this value is 72.

The m/z value is the mass/charge ratio of ions formed in the mass spectrometer. As these ions are nearly always unipositive, the value is often equal to the mass of the ion formed.

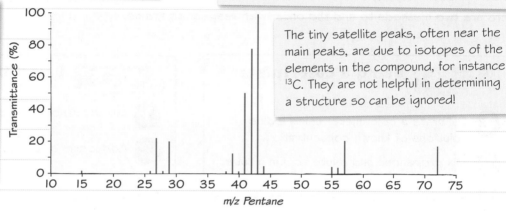

The tiny satellite peaks, often near the main peaks, are due to isotopes of the elements in the compound, for instance ^{13}C. They are not helpful in determining a structure so can be ignored!

The other peaks form when the energy in the mass spectrometer causes the molecular ion to split and form other ionic fragments. The peaks at 57 and 43 are due to $CH_3CH_2CH_2CH_2^+$ and $CH_3CH_2CH_2^+$ ions.

The equations for the formation of the fragments at 57 and 43 are:

$C_5H_{12}^+ \rightarrow C_4H_9^+ + CH_3$ and $C_5H_{12}^+ \rightarrow C_3H_7^+ + C_2H_5$

Fragmentation always results in the formation of a new ion and a neutral species. The neutral particle is not detected by the mass spectrometer.

Worked example

Write an equation for the formation of the fragment shown at $\frac{m}{z} = 29$ on the mass spectrum of pentane.
(2 marks)

$C_5H_{12}^+ \rightarrow C_2H_5^+ + C_3H_7$

The parent ion, the fragment ion **and** the neutral fragment would all be needed in the equation.

A common pattern seen in the mass spectrum of alkanes shows peaks at $\frac{m}{z} = 15, 29, 43$ and 57 respectively. Explain this pattern. **(2 marks)**

This pattern arises when successive ionic fragments, as $-CH_2^-$, split off from the molecular ion as the $\frac{m}{z}$ values of each peak differ by 14. For example $\frac{m}{z} = 15$ is CH_3^+, $\frac{m}{z} = 29$ is $CH_3CH_2^+$ and so on.

Now try this

The mass spectrum shown is for the alkene hept-1-ene. Identify the ions that cause peaks A, B and C. Write an equation for the formation of each ion. Peak A, $\frac{m}{z} = 27$; Peak B, $\frac{m}{z} = 41$; Peak C, $\frac{m}{z} = 98$
(6 marks)

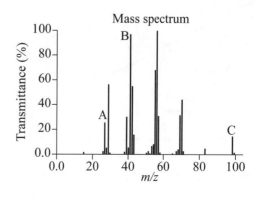

Concentration–time graphs (zero order reactants)

If you can measure the concentration of a reactant or product during a reaction, then you can plot a concentration–time graph. From this graph, you can deduce the **order** with respect to a reactant.

The iodine/thiosulfate reaction

Iodine reacts with propanone, with an acid catalyst:

$$CH_3COCH_3 + I_2 \rightarrow CH_3COCH_2I + HI$$

Iodine solution is coloured and it reacts with sodium thiosulfate solution.

Here are two methods to find the order with respect to iodine.

> The concentrations of the propanone and of the acid are much larger than that of the iodine. These concentrations will change very little during the reaction so will not affect the rate of reaction during the experiment.

Practical skills — Using a colorimeter

1. Calibrate a colorimeter using iodine solutions of known concentration.

2. Mix propanone and dilute acid in a tube.

3. Add iodine solution, start timing, and mix.

4. Place the tube in the colorimeter.

5. At fixed time intervals, take a reading of the absorbance from the colorimeter.

6. Use the absorbance values and the calibration curve to calculate the concentration of iodine.

Practical skills — Using titration

1. Mix propanone and dilute acid in a flask.

2. Add iodine solution, start timing, and mix.

3. At fixed time intervals, take a sample of the reaction mixture using a pipette.

4. Quench the reaction by adding the sample to some sodium hydrogencarbonate solution.

5. Titrate the sample using sodium thiosulfate.

6. Use the titration values to calculate the concentration of iodine.

> The sodium hydrogencarbonate neutralises the acid catalyst and stops the reaction.

Concentration–time graph: zero order

The results of either experiment can be plotted.

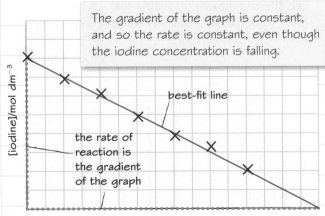

> The gradient of the graph is constant, and so the rate is constant, even though the iodine concentration is falling.

best-fit line

the rate of reaction is the gradient of the graph

[iodine]/mol dm^{-3}

Time/minutes

When the rate of reaction does not depend on the concentration of a reactant, the reaction is **zero order** with respect to that reactant. The concentration–time graph is a straight line.

Worked example

Zinc reacts with copper(II) sulfate solution.

Which apparatus could be used to determine the effect of the concentration of $CuSO_4$(aq) on the rate of reaction? **(1 mark)**

A balance
B gas syringe
C colorimeter
D pH meter

C

> No gas is formed, so no change in mass (A, B incorrect). No H^+ or OH^- ions (D incorrect). $CuSO_4$(aq) is blue and $ZnSO_4$(aq) is colourless so **C** is correct.

Now try this

Explain what is meant by *zero order* for a reactant. Describe in outline how you would carry out an experiment to show that one of the reactants was zero order. **(3 marks)**

Concentration–time graphs (first order reactants)

When the rate of reaction is proportional to the concentration of a reactant, the reaction is **first order** with respect to that reactant. The concentration–time graph is a curve.

Decomposition of dinitrogen pentoxide

An example of a first order reaction is the decomposition of N_2O_5 to form nitrogen dioxide and oxygen.

$2N_2O_5(aq) \rightarrow 4NO_2(aq) + O_2(g)$

The concentration of the dinitrogen pentoxide can be measured as the reaction occurs.

Concentration–time graph: first order

The graph shows the results of an experiment to monitor the decomposition of N_2O_5.

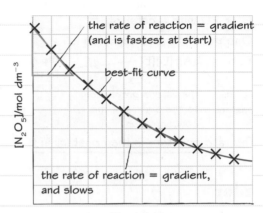

the rate of reaction = gradient (and is fastest at start)

best-fit curve

the rate of reaction = gradient, and slows

$[N_2O_5]/mol\ dm^{-3}$

Time/minutes

Half-life

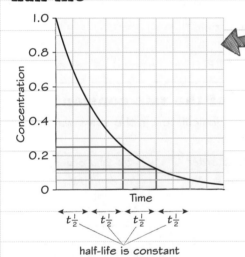

Concentration

Time

$t\frac{1}{2}$ $t\frac{1}{2}$ $t\frac{1}{2}$ $t\frac{1}{2}$

half-life is constant

In a reaction:

- The half-life is the time taken for the concentration of a reactant to halve.
- For a first order reaction, this value is a constant.
- The half-life can be measured from the concentration–time graph.
- If the half-life measured at different times on a concentration–time graph is constant, then the reaction is first order.

In a reaction, the time taken for the concentration of reactant A, [A], to fall is measured.

[A]/mol dm⁻³	Time/s
2.0	0
1.5	31
1.0	82
0.5	165

Use evidence from the results to show that the reaction is first order with respect to A. **(2 marks)**

From 2.0 to 1.0 mol dm⁻³, the half-life is 82 s.

From 1.0 to 0.5 mol dm⁻³, the half-life is 83 s.

The half-life is approximately constant so the reaction is first order with respect to A.

Gentian violet (a dye) reacts with hydroxide ions to give a colourless product.

Outline a practical method to show that the reaction is first order with respect to hydroxide ions.

(4 marks)

Calculate the length of time for the concentration to fall by half; so from 1.0 to 0.5 mol dm⁻³ the time is 165 − 82 = 83s.

As these are experimental results, it is fair to say that 82s and 83s are approximately equal.

Rate equation and rate constant

The rate of a chemical reaction depends on the concentration of some or all of the reactants and the catalyst. This relationship is shown in the rate equation.

Maths skills — A rate equation

Consider the reaction A + 2B + C → 2D with a catalyst, E.

One possible rate equation is

Rate = $k[A]^2[B][E]$ ——— the rate is proportional to the concentrations of B and the catalyst

the rate constant — the rate is proportional to the square of the concentration of A

The rate constant is the constant in the rate equation.

Some key points

✓ The rate equation is found by experiment and cannot be written down from the chemical equation.

✓ [A] means the concentration of A, measured in mol dm^{-3}.

✓ Some or all of the reactants may be in the rate equation (see page 82 for the reason), and the catalyst concentration may be in the rate equation.

Order

The order of a reactant is the power to which the reactant's (or catalyst's) concentration is raised in the rate equation.

In the example above:

Order with respect to A = 2

Order with respect to B = 1

Order with respect to E = 1

C does not appear in the rate equation. The order with respect to C is zero order.

Overall order

The overall order of a reaction is the sum of the individual orders in the rate equation. In the example above:

overall order = 2 + 1 + 1 = 4

The rate constant

This can be found from experimental data.

1 For example, if rate = $k[A][B]$

and rate = 0.80 mol dm^{-3} s^{-1} when [A] = 0.15 mol dm^{-3} and [B] = 0.050 mol dm^{-3}

Then $k = \dfrac{\text{rate}}{[A][B]} = \dfrac{0.800}{(0.15 \times 0.050)}$ = 107 dm^3 mol^{-1} s^{-1}

2 For a first order reaction, $k = \dfrac{\ln 2}{t_{\frac{1}{2}}}$

3 For a first order reaction k is the gradient of the rate-concentration graph (see page 79)

Maths skills — Units of the rate constant

Units are calculated in the same way as the value.

Zero order	First order	Second order
rate = k	rate = $k[A]$	rate = $k[A]^2$
k = rate	$k = \dfrac{\text{rate}}{[A]}$	$k = \dfrac{\text{rate}}{[A]^2}$
k = mol dm^{-3} s^{-1}	$k = \dfrac{\text{mol dm}^{-3}\text{ s}^{-1}}{\text{mol dm}^{-3}}$	$k = \dfrac{\text{mol dm}^{-3}\text{ s}^{-1}}{\text{mol}^2\text{ dm}^{-6}}$
	= s^{-1}	= dm^3 mol^{-1} s^{-1}

Maths skills

ln2 is the natural logarithm of 2. Press LN then 2 then = on your calculator (= 0.693). Do not use LOG.

Worked example

The rate equation for the reaction between A and B is rate = $k[A][B]^2$

[A] / mol dm^{-3}	[B] / mol dm^{-3}	Rate / mol dm^{-3}
0.020	0.020	1.2×10^{-4}

Use the information to give the order with respect to A and B, and to find k.   **(4 marks)**

Order with respect to A = 1, and B = 2.

$k = \dfrac{1.2 \times 10^{-4}}{(0.020)(0.020)^2}$ = 15 dm^6 mol^{-2} s^{-1}

Now try this

Use the rate equation and the value of k from the worked example to find the rate of reaction when [A] = 0.050 mol dm^{-3} and [B] = 0.080 mol dm^{-3}

(2 marks)

Finding the order

The order of a reactant in the rate equation can be deduced from experimental data, or from the shape of a rate-concentration graph (see also pages 78 and 79).

Finding orders from initial rate data

The initial rate of reaction between substances A and B is shown.

Experiment	[A]/mol dm^{-3}	[B]/mol dm^{-3}	Initial rate/mol dm^{-3} s^{-1}
1	0.01	0.20	0.10
2	0.02	0.20	0.20
3	0.01	0.40	0.40

[B] the same, [A] × 2, rate × 2 : first order with respect to A

[A] the same, [B] × 2, rate × 4 : second order with respect to B

The rate equation is rate = k[A][B]^2

0 The order is zero if a change in concentration of a reactant has no effect on the rate.

1 The order is first if the change in rate is the same as the change in concentration.

2 The order is second if the change in rate is the square of the change in concentration.

The initial rate is used (the rate at the beginning of the experiment) because we know the concentrations of the reactants at this point.

Maths skills — Finding order from a rate–concentration graph

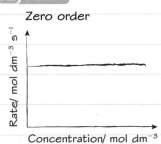

Zero order

The concentration has no effect on the rate, giving a graph with constant rate.

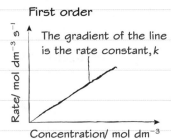

First order

The gradient of the line is the rate constant, k

The rate is proportional to the concentration giving a graph with a straight line through the origin.

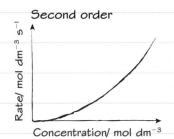

Second order

The rate is proportional to the square of the concentration giving a graph with a curve.

Worked example

For the reaction $PCl_3 + Cl_2 \rightarrow PCl_5$, the following data were obtained:

Experiment	[PCl$_3$] / mol dm^{-3}	[Cl$_2$] / mol dm^{-3}	Rate mol / dm^{-3} s^{-1}
1	0.36	1.26	6.0×10^{-4}
2	0.36	0.63	1.5×10^{-4}
3	0.72	2.52	4.8×10^{-3}

Find the order with respect to each reactant and write down the rate equation. **(3 marks)**

Experiments 2 and 1: [PCl$_3$] same, [Cl$_2$] × 2, rate × 4: second order with respect to Cl$_2$.

Experiments 1 and 3: [PCl$_3$] × 2, [Cl$_2$] × 2, rate × 8.

As Cl$_2$ is second order, doubling the chlorine concentration will increase the rate by 4.

So, doubling the PCl$_3$ concentration had the effect of doubling the rate (to overall give rate × 8). So, first order with respect to PCl$_3$. Rate = k[PCl$_3$][Cl$_2$]^2

Now try this

The following rate data were collected for the reaction:
$$2NO(g) + H_2(g) \rightarrow N_2O(g) + H_2O(g)$$

Find the order with respect to each reactant and write down the rate equation. **(3 marks**

[NO] / mol dm^{-3}	[H$_2$] / mol dm^{-3}	Rate / mol dm^{-3}
0.60	0.37	3.0×10^{-3}
1.20	0.37	1.2×10^{-2}
1.20	0.74	1.2×10^{-2}

The rate-determining step

The rate-determining step is the slow step in a multi-step reaction, and the reactants in this step appear in the rate equation. The series of steps by which a reaction occurs is called the **mechanism**.

Deducing the rate-determining step from the rate equation

For the reaction

$NO_2 + CO \rightarrow NO + CO_2$

rate = $k[NO_2]^2$

The rate equation is used to deduce the steps in the reaction:

Step I: $2NO_2 \rightarrow NO + NO_3$

Step 2: $NO_3 + CO \rightarrow NO_2 + CO_2$

- The rate equation suggests that two molecules of NO_2 (because the order is 2) are in the slow step.
- No molecules of CO (because the order is 0) are in the slow step.
- Step I must be the slow step (rate-determining step).

One step reaction

If the reaction on the left occurred in only one step, then the rate equation would be:

rate = $k[NO_2][CO]$

because one molecule of NO_2 and one molecule of CO would be in the slow (only) step.

$2NO_2 + \cancel{NO_3} + CO$
$\rightarrow NO + \cancel{NO_3} + \cancel{NO_2} + CO_2$

If the two steps are added together then the overall equation is found.

Nucleophilic substitution in tertiary haloalkanes

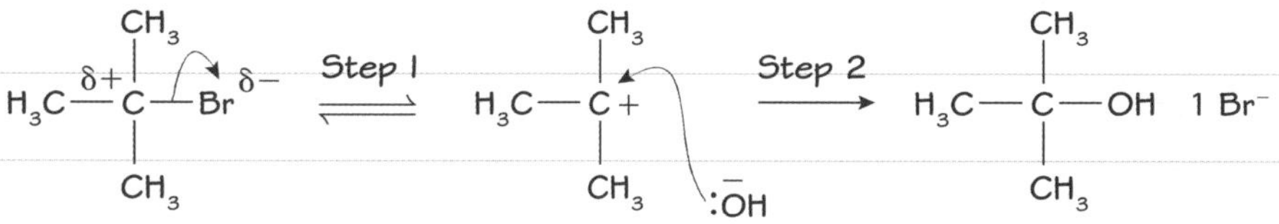

The S_N1 mechanism is:

- Rate equation is rate = $k[(CH_3)_3CBr]$
- Step I is the slow step.
- Only the haloalkane is in the slow step.

Nucleophilic substitution in primary haloalkanes

The S_N1 mechanism is:

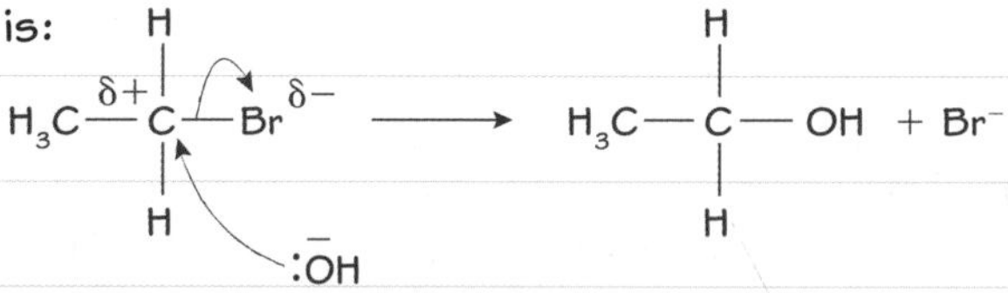

- The rate equation is rate = $k[CH_3CH_2Br][OH^-]$
- There is only one step.
- The haloalkane and a hydroxide ion are in this step.

Be careful!

In S_N1 and S_N2 mechanisms, 'I' and '2' indicate the order, not the number of steps.

More details about haloalkane substitution with the reason for these different mechanisms is on page 70.

Worked example

Bromine can be formed by the oxidation of HBr.

Step 1: $HBr + O_2 \rightarrow HBrO_2$

Step 2: $HBrO_2 + HBr \rightarrow 2HBrO$

Step 3: $HBrO + HBr \rightarrow Br_2 + H_2O$

The rate equation is rate = $k[HBr][O_2]$. Explain which step is the slow step.   **(2 marks)**

Step I. This step involves one HBr molecule and one O_2 molecule.

Now try this

1 Give the orders with respect to HBr and O_2 in the reaction shown on the left and the overall order for the reaction.   **(3 marks)**
2 Write the overall equation for the reaction.   **(1 mark)**

Make sure that intermediates (in this case $HBrO_2$ and HBrO) cancel out. You may have to multiply one or all of the steps to make sure that intermediates made in one step are used up in the next.

The Arrhenius equation

As temperature increases, the rate of reaction increases, so the rate constant for a reaction will be higher at a higher temperature.

The Arrhenius equation

The relationship between the rate constant and temperature is given by the Arrhenius equation.

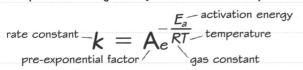

rate constant — $k = Ae^{-\frac{E_a}{RT}}$ — temperature

pre-exponential factor ∕   gas constant ∖ activation energy

The Arrhenius equation is given on the Data Sheet in your exam.

The rate constant varies with temperature

As the temperature increases, the rate constant increases.

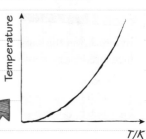

as $T \uparrow$, $\frac{E_a}{RT} \downarrow$, $e^{-\frac{E_a}{RT}} \uparrow$, $k \uparrow$

Practical skills   Finding an activation energy using an iodine clock

Hydrogen peroxide reacts with iodide ions in an acidic solution to make iodine:

$H_2O_2(aq) + 2I^-(aq) + 2H^+(aq) \rightarrow I_2(aq) + 2H_2O(l)$

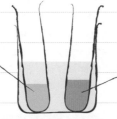

mixture of potassium iodide and sodium thiosulfate

mixture of hydrogen peroxide, dilute acid and starch

Water is heated to the required temperature and the contents of the two boiling tubes.

Some sodium thiosulfate is added which reacts with the iodine formed. When all of the sodium thiosulfate is reacted, iodine remains. Starch is added, so a blue-black colour is then seen.

The experiment is repeated at different temperatures.

Maths skills 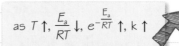 Taking logs of the Arrhenius equation gives $\ln k = \frac{E_a}{RT} + \ln A$

A graph is plotted of $\ln(\frac{1}{time})$ on the y axis against $\frac{1}{T}$ on the x-axis.

The activation energy is calculated using:

$E_a = -gradient \times R$

gradient $= -\frac{E_a}{R}$

in a clock reaction, $1/time$ is proportional to k

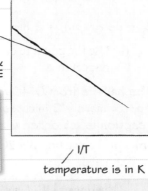

Be careful! Note that the gradient of this graph is negative, giving a positive value for E_a.

temperature is in K

1 Add the contents of one boiling tube to the other and leave in the water bath, start timing.

2 Stop the clock when a blue-black colour appears.

Worked example

A graph of results, shown above, is plotted for a clock reaction forming iodine.

(a) How can A, the pre-exponential factor, be found from the graph? **(1 mark)**

The intercept on the y-axis is $\ln A$.

(b) Explain why $\frac{1}{time} \propto k$. **(2 marks)**

Average rate $= \dfrac{\text{change in concentration}}{\text{time}}$

A fixed amount of iodine is formed in each run, so the change in concentration is constant. Average rate, and k, are proportional to $\frac{1}{t}$.

Now try this

The graph shows the relationship between the logarithm of the rate constant of a reaction, $\ln k$, and the reciprocal of the temperature, $\frac{1}{T}$.

What value is given by the gradient? **(1 mark)**

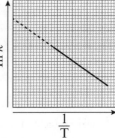

A E_a **B** $-E_a$ **C** $-\dfrac{E_a}{R}$ **D** $\dfrac{E_a}{R}$

Exam skills 7

This exam-style question uses knowledge and skills you have already revised. Look at pages 78–82 for a reminder about the rate equation and order, half-life and the rate-determining step.

Worked example

In a reaction the substance A decomposes. The concentration of A is measured every minute for 10 minutes. The results are plotted on the graph.

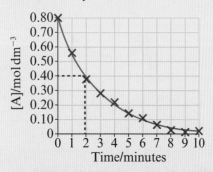

(a) (i) Find the half-life of the reaction from the initial [A]. **(2 marks)**

Initial [A] = 0.80 mol dm⁻³

$$\frac{0.80}{2} = 0.40 \text{ mol dm}^{-3}$$

$t_{\frac{1}{2}}$ (see graph) = 2 minutes

(ii) Use the graph to deduce the order of the reaction with respect to A. **(2 marks)**

The half-life from 0.40 mol dm⁻³ to 0.20 mol dm⁻³ is also approximately 2 minutes. A constant half-life means that the reaction is 1st order with respect to A.

(b) For the reaction B + 2C → D, the following initial rate data was gathered.

Experiment	[B] / mol dm⁻³	[C] / mol dm⁻³	Rate / mol dm⁻³ s⁻¹
1	0.36	1.26	6.0 × 10⁻⁴
2	0.36	0.63	1.5 × 10⁻⁴
3	0.72	2.52	4.8 × 10⁻³

(i) Deduce the orders with respect to B and C. **(2 marks)**

Experiments I and 2:

[B] same, [C] × $\frac{1}{2}$, rate × $\frac{1}{4}$.

rate ∝ [C]² so **second order** with respect to C.

Experiments I and 3:

[B] × 2, [C] × 2, rate × 8.

[C] × 2 will give rate × 4 (as second order), so [B] × 2 has given rate × 2.

rate ∝ [B] **first order** with respect to B.

(ii) Suggest, and explain, how many steps are found in the mechanism of this reaction. **(1 mark)**

One: as the orders tell us that one B particle and two C particles are involved in the slow step, the slow step could be B + 2C → D, so no other steps are required.

🖩 **Maths skills** When you are extracting information from a graph (either given or one you have drawn), read the graph carefully noting the scale. In this case, you need to halve the initial concentration. Then use a sharp pencil and ruler to find the time taken for the initial concentration to halve.

🖩 **Maths skills** To check whether a half-life is constant, you can use **any** starting point. In the answer shown, the time for the concentration to fall from 0.40 mol dm⁻³ to 0.20 mol dm⁻³ has been used, but 0.40 mol dm⁻³ is only one possible value to halve. (However, it is wise to choose a number easy to halve and easy to read from the graph scale.)

To find the first of the orders, choose a pair of experiments where the concentrations of all of the reactants except one remain constant, so that the effect of changing the remaining reactant can be seen.

Remember that the order usually tells us the number of particles of that species in the slow step (rate-determining step).

Finding the equilibrium constant

The equilibrium constant, introduced on page 53, can be calculated by finding, by experiment, the concentration of each substance at equilibrium.

 **Practical skills** — **Ester hydrolysis**

The equation for an ester hydrolysis is:

$$CH_3COOC_2H_5(aq) + H_2O(l)$$
$$\rightleftharpoons CH_3COOH(aq) + C_2H_5OH(aq)$$

Method

1. Pipette dilute hydrochloric acid (the catalyst) to several dry, stoppered boiling tubes.

2. Weigh each tube.

3. Using a dry measuring cylinder, add a measured amount of ethyl ethanoate to half of the tubes.

4. Reweigh each tube.

5. Mix the solutions and then leave for several days.

6. Pour each mixture into a small flask and then titrate with sodium hydroxide solution of known concentration.

Experimental notes

- A pipette is used for accuracy in 1. Several tubes are set up to allow the reliability of the results to be checked.

- The masses are used in 2 and 4 to calculate the mass of ethyl ethanoate added.

- The reaction is slow at 5, so the mixtures are left to allow the reaction to reach equilibrium.

- At 6 the sodium hydroxide solution reacts with the acid catalyst and with the ethanoic acid formed in the reaction.

Analysing the results

1. The titration results for the boiling tubes containing only acid can be used to calculate the average moles of acid catalyst.

2. The results for the other tubes can be used to calculate the average total moles of acid catalyst + ethanoic acid.

3. Moles of ethanoic acid formed = moles from 2 − average moles from 1.

4. The moles of ethanol formed = moles of ethanoic acid formed (1:1 ratio in equation).

5. The average moles of ethyl ethanoate at the start is calculated from the mass added and ethyl ethanoate's density. As 1 mol of ethyl ethanoate forms 1 mol of ethanol, the moles of ethyl ethanoate at equilibrium = average moles at start − moles of ethanol formed from 4.

6. The average moles of water at the start is calculated from the mass of acid catalyst added (almost all of the mass of the dilute acid solution is water). As 1 mol of water forms 1 mol of ethanol, the moles of water at equilibrium = average moles at start − moles of ethanol formed from 4.

7. K_c can then be calculated using:

$$K_c = \frac{[\text{ethanol}][\text{ethanoic acid}]}{[\text{ethyl ethanoate}][\text{water}]}$$

Worked example

Ethyl ethanoate (2.0 mol) and water (2.0 mol) are mixed with a catalyst and left to reach equilibrium. At equilibrium, 0.65 mol ethanol has formed.

Find K_c.   **(1 mark)**

$$K_c = \frac{[\text{ethanol}][\text{ethanoic acid}]}{[\text{ethyl ethanoate}][\text{water}]}$$

$$= \frac{(0.65 / V)(0.65 / V)}{(1.35 / V)(1.35 / V)}$$

$$= 0.23$$

Now try this

Iodine (3.0 mol) and hydrogen (3.6 mol) were mixed and left to form an equilibrium. $H_2 + I_2 \rightleftharpoons 2HI$
At equilibrium, there was 2.4 mol HI. Find K_c.

(4 marks)

Take great care working out the amount of each substance at equilibrium. They go down (reactants) and up (products) in the same ratio as in the equation.

Calculating the equilibrium constant, K_c

The equilibrium constant, K_c, can be calculated for homogeneous and heterogeneous equilibria.

A homogeneous equilibrium

$CO(g) + 2H_2(g) \rightleftharpoons CH_3OH(g)$

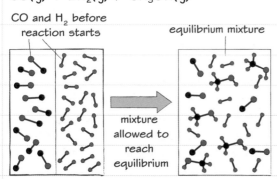

CO and H_2 before reaction starts

equilibrium mixture

mixture allowed to reach equilibrium

All substances completely mixed.

A heterogeneous equilibrium

$Cu(s) + 2Ag^+(aq) \rightleftharpoons Cu^{2+}(aq) + 2Ag(s)$

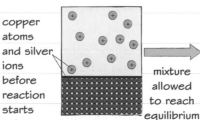

copper atoms and silver ions before reaction starts

mixture allowed to reach equilibrium

aqueous part of equilibrium mixture

solid part of equilibrium mixture

Substances in two phases, solid and aqueous.

Substances with different states in the equilibrium constant

The concentrations of liquid and solid substances are omitted from equilibrium constants.
Only the concentrations of gases and solutions are included.

K_c for a homogeneous equilibrium

In a homogeneous equilibrium mixture, all substances are in the same phase.

For the reaction above,

$$K_c = \frac{[CH_3OH]}{[CO][H_2]^2}$$

 The power of each substance in the K_c expression is the number of moles of that substance in the balanced equation.

K_c for a heterogeneous equilibrium

In a heterogeneous equilibrium, the substances are in different phases (they are not completely mixed).

For the reaction above,

$$K_c = \frac{[Cu^{2+}]}{[Ag^+]^2}$$

 Cu(s) and Ag(s) are omitted from K_c because they are solids.

Worked example

A mixture of CO and H_2 is used to make methanol.
The mixture was allowed to reach equilibrium in a 2.0 dm³ sealed vessel.

$$CO(g) + 2H_2(g) \rightleftharpoons CH_3OH(g)$$

The number of moles of each component at equilibrium is shown below.

CO(g)	H₂(g)	CH₃OH(g)
6.20×10^{-3}	4.80×10^{-2}	5.20×10^{-5}

The number of moles is divided by 2, the volume, to convert to concentration.

Calculate K_c with units.

$$K_c = \frac{[CH_3OH]}{[CO][H_2]^2} = \left(\frac{5.2 \times 10^{-5}}{2}\right) \bigg/ \left(\frac{6.2 \times 10^{-3}}{2}\right)\left(\frac{4.8 \times 10^{-2}}{2}\right)^2$$

$$= 14.6 \text{ dm}^6 \text{ mol}^{-2}$$

⊞ Maths skills — Units of K_c

Units are calculated in the same way as the value.
In the left-hand example:

$$K_c = \frac{[CH_3OH]}{[CO][H_2]^2}$$

$$= \frac{\text{mol dm}^{-3}}{(\text{mol dm}^{-3})(\text{mol dm}^{-3})^2}$$

$$= \text{dm}^6 \text{ mol}^{-2}$$

In the right-hand example:

$$K_c = \frac{[Cu^{2+}]}{[Ag^+]^2} = \frac{\text{mol dm}^{-3}}{(\text{mol dm}^{-3})^2}$$

$$= \text{dm}^3 \text{ mol}^{-1}$$

Now try this

Write expressions for K_c and give the units for:
(a) $2SO_2(g) + O_2(g) \rightleftharpoons 2SO_3(g)$ (b) $N_2O_4(g) \rightleftharpoons 2NO_2(g)$ (c) $Zn(s) + Cu^{2+}(aq) \rightleftharpoons Zn^{2+}(aq) + Cu(s)$ **(6 marks)**

Calculating K_p

For an equilibrium involving gases, the equilibrium constant can be expressed in terms of the **partial pressures** of the reactants and products.

Partial pressure in a mixture of gases

$$CO(g) + 2H_2(g) \rightleftharpoons CH_3OH(g)$$

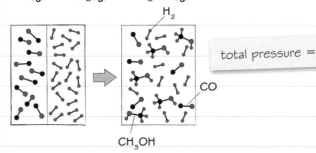

total pressure = p

The partial pressure of each of these gases is the pressure that each gas would exert if it occupied the container alone.

Mole fraction and partial pressure

Mole fraction of a gas, $x = \dfrac{\text{moles of the gas}}{\text{total moles of gas}}$

In the diagram, left:

$$x(H_2) = \frac{10}{20} = 0.5$$

$$x(CO) = \frac{5}{20} = 0.25$$

$$x(CH_3OH) = \frac{5}{20} = 0.25$$

$$x(H_2) + x(CO) + x(CH_3OH) = 1$$

The partial pressure = mole fraction × total pressure

$$p(H_2) = 0.5p \qquad p(CO) = 0.25p$$
$$p(CH_3OH) = 0.25p$$

Calculating K_p

The expression for K_p is constructed in a similar way to K_c, except that partial pressures are used instead of concentrations.

In the example above, $K_p = \dfrac{p(CH_3OH)}{p(CO)p(H_2)^2}$

 Maths skills **Units of K_p**

Units are calculated in the same way as the value. Partial pressures will be given in Pa, kPa or atm.

In the example:

$$K_p = \frac{p(CH_3OH)}{p(CO)p(H_2)^2} = \frac{Pa}{(Pa)(Pa^2)} = Pa^{-2}$$

Homogeneous and heterogeneous equilibria

 K_p is only used for equilibria with gaseous reactants and/or products.

 In a homogeneous equilibrium of all gases, all substances are included in K_p.

In a heterogeneous equilibrium with gases and some solids and/or liquids, the solid and/ or liquid substances are omitted from K_p.

Worked example

In the reaction above, at equilibrium there are 0.94 mol CO, 0.20 mol H_2 and 0.06 mol of methanol. The total pressure is 1×10^5 Pa. Calculate K_p.   **(4 marks)**

Total moles = 1.20

$$p(CO) = \frac{0.94}{1.20} \times 1 \times 10^5 = 7.83 \times 10^4 \text{ Pa}$$

$$p(H_2) = \frac{0.20}{1.20} \times 1 \times 10^5 = 1.67 \times 10^4 \text{ Pa}$$

$$p(CH_3OH) = \frac{0.06}{1.20} \times 1 \times 10^5 = 5.00 \times 10^3 \text{ Pa}$$

$$K_p = \frac{p(CH_3OH)}{p(CO)p(H_2)^2} = \frac{5.00 \times 10^3}{(7.83 \times 10^4)(1.67 \times 10^4)^2}$$

$$= 2.29 \times 10^{-10} \text{ Pa}^{-2}$$

A good check is to see if your partial pressures add up to the total pressure.

Now try this

Sulfur dioxide was mixed with oxygen in a sealed flask.

Equilibrium was established according to the equation $2SO_2(g) + O_2(g) \rightleftharpoons 2SO_3(g)$

At equilibrium, there were 0.75 mol SO_2, 0.18 mol O_2 and 7.5 mol SO_3, with a total pressure of 1200 kPa.

(a) Calculate the mole fraction of each gas.
   (1 mark)

A good check is to see if your mole fractions add up to 1.

(b) Calculate the partial pressure of each gas.
   (1 mark)

(c) Write an expression for K_p and calculate its value, including a unit.   **(3 marks)**

The equilibrium constant under different conditions

If the conditions of a reaction at equilibrium are altered, then the position of equilibrium may alter, but the equilibrium constant is only different at a different temperature.

Equilibrium constant and temperature

1 For an exothermic forward reaction, Le Chatelier's principle lets us predict that the position of equilibrium will shift left (the exothermic direction) as the temperature is raised (see page 52). K_c and K_p have a **lower** value at a higher temperature.

2 For an endothermic forward reaction the equilibrium will shift right (the endothermic direction) as the temperature is raised. K_c and K_p have a **higher** value at a higher temperature.

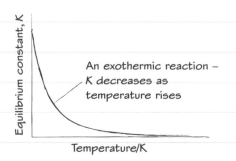

An exothermic reaction – K decreases as temperature rises

- $[H_2O]$ is omitted from K_W and K_a
- The concepts that apply to equilibria also apply to K_W and K_a

Equilibrium constants in acids

In the acids topic you will find two specific equilibrium constants.

Water forms a small number of ions.

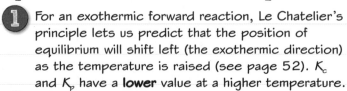

$$2 H_2O \rightleftharpoons H_3O^+ + OH^-$$
$$K_w = [H_3O^+][OH^-]$$

This equilibrium constant is called the ionic product (see page 91).

Weak acids (HA) partially dissociate to form ions.

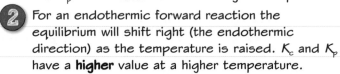

$$HA + H_2O \rightleftharpoons H_3O^+ + A^-$$
$$K_a = \frac{[H_3O^+][A^-]}{[HA]}$$

This equilibrium constant is called the acid dissociation constant, K_a (see page 92).

Equilibrium constant and changes in conditions

- Altering the concentration or pressure may shift the position of equilibrium (see page 52). However, the equilibrium constant is unchanged.

- Adding a catalyst does not shift the position of equilibrium (see page 52) and the equilibrium constant is unchanged.

Worked example

Consider the equilibrium

$$CO(g) + H_2O(g) \rightleftharpoons CO_2(g) + H_2(g)$$

Explain, if the pressure is increased,
(a) the effect on the position of equilibrium. **(2 marks)**

There are 2 gas moles on each side of the equation, so changing the pressure has no effect.

(b) the effect on K_p. **(1 mark)**

There is no change because only a change in temperature affects K_p.

Now try this

Some students calculated the value of K_c for
HCOOH + CH₃OH ⇌ HCOOCH₃ + H₂O
by titrating the equilibrium mixture with NaOH.
Their K_c values were different.
Which reason(s) below could explain this? **(1 mark)**
Each student's experiment:
A was at a different temperature
B used a different concentration of NaOH
C titrated a different volume of the equilibrium mixture.

Brønsted–Lowry acids and bases

Brønsted–Lowry acids are proton donors and Brønsted-Lowry bases are proton acceptors.

A proton transfer reaction

$NH_3 + HCl \rightarrow NH_4^+ + Cl^-$

A hydrogen atom has one proton and one electron. A hydrogen ion, H^+, is a proton.

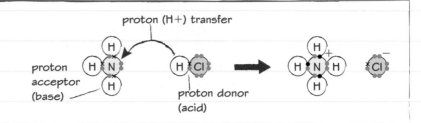

proton (H+) transfer

proton acceptor (base)

proton donor (acid)

Conjugate acid base pairs

Carbonate ions react in water:

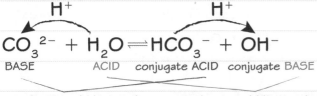

$$CO_3^{2-} + H_2O \rightleftharpoons HCO_3^- + OH^-$$

BASE   ACID   conjugate ACID   conjugate BASE

conjugate acid-base pair   conjugate acid-base pair

An acid donates a proton; the resultant species is the **conjugate base**.

In the reverse reaction, this conjugate base can regain a proton to reform the acid.

A base particle accepts a proton; the resultant species is the **conjugate acid**.

In the reverse reaction, this conjugate acid can donate a proton to reform the base.

Monobasic, dibasic and tribasic acids

Monobasic: $HCl(aq) + OH^-(aq) \rightarrow H_2O(l) + Cl^-(aq)$

Dibasic: $H_2SO_4(aq) + 2OH^-(aq) \rightarrow 2H_2O(l) + SO_4^{2-}$ (aq)

Tribasic: $H_3PO_4(aq) + 3OH^-(aq) \rightarrow 3H_2O(l) + PO_4^{3-}$ (aq)

A monobasic acid molecule donates one H^+ ion, a dibasic acid two H^+ ions and a tribasic acid three H^+ ions.

Dissociation

✓ The first dissociation of strong acids, like those above, to give H_3O^+ ions, is almost complete.

✓ The second (H_2SO_4 and H_3PO_4) and third (H_3PO_4) dissociations of these acids is much less complete.

✓ Weak acids (for example, carboxylic acids) only slightly dissociate.

Ionic equations for acid reactions

In general acid reactions, the H^+ ions react to give standard products.

✓ metals [→ hydrogen]

$M + nH^+ \rightarrow M^{n+} + \frac{n}{2}H_2$

✓ carbonates [→ carbon dioxide and water]

$CO_3^{2-} + 2H^+ \rightarrow CO_2 + H_2O$

✓ metal oxides [→ water]

$MO_{\frac{n}{2}} + nH^+ \rightarrow M^{n+} + \frac{n}{2}H_2O$

✓ alkalis [→ water]

$OH^- + H^+ \rightarrow H_2O$

Acids whose (first) dissociation to give H_3O^+ ions is (almost) complete are **strong acids** (for example, HCl, H_2SO_4). Acids where dissociation is only slight are **weak acids** (for example, carboxylic acid).

Worked example

Consider the reaction:

$HCl + H_2PO_4^- \rightarrow Cl^- + H_3PO_4$

(a) Identify the acid–base conjugate pairs. (**3 marks**)

HCl is the acid and Cl^- is the conjugate base.

$H_2PO_4^-$ is the base and H_3PO_4 is the conjugate acid.

Now try this

In the nitration of benzene, concentrated nitric and sulfuric acids are mixed.

They react as follows:

$HNO_3 + 2H_2SO_4 \rightleftharpoons NO_2^+ + H_3O^+ + 2HSO_4^-$

Identify the acid–base conjugate pairs.   (**3 marks**)

pH

The pH of a solution is a measure related to the concentration of H^+ ions.

The pH scale

0 1 2 3 4 5 6 7 8 9 10 11 12 13 14

more acidic ◀──────────▶ more alkaline

solutions of increasing $[H^+]$ solutions of reducing $[H^+]$

pH

$pH = -\log[H^+]$
Note that pH has no units.

Maths skills Type – then LOG then the concentration of H^+ ions then =

Maths skills To find $[H^+]$ from the pH value, use:
$[H^+] = 10^{-pH}$ 10^x is the inverse of log.

When an acid dissociates, it donates an H^+ ion to a water molecule, forming H_3O^+. However, in the pH equation this species is shown as H^+

Worked example

pH when a strong acid is partly neutralised

Calculate the pH of the mixture formed when 25 cm³ of 0.10 mol dm⁻³ HCl has 40 cm³ of 0.050 mol dm⁻³ KOH added. **(3 marks)**

$HCl + KOH \rightarrow KCl + H_2O$

original mol HCl $= \dfrac{25}{1000} \times 0.10$

$= 0.0025$ mol

original mol KOH $= \dfrac{40}{1000} \times 0.050$

$= 0.0020$ mol

After neutralisation:

mol acid $= 0.0025 - 0.0020$

$= 0.00050$ mol

$[H^+] = \dfrac{0.0050}{0.065}$

$= 0.00769$ mol dm⁻³

$pH = -\log(0.00769) = 2.1$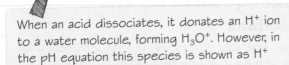

The volume is the total volume,
$25 + 40 = 65$ cm³
$= 0.065$ dm³

Worked example

pH of strong acids

(a) Calculate the pH of 0.010 mol dm⁻³ HCl. **(1 mark)**

$HCl + H_2O \rightarrow H_3O^+ + Cl^-$
Monobasic acid, so $[H^+] = [HCl]$
$= 0.010$ mol dm⁻³
$pH = -\log(0.01) = 2.0$

(b) Calculate the pH of 0.10 mol dm⁻³ H_2SO_4, assuming that the acid is fully dissociated. **(2 marks)**

$H_2SO_4 + 2H_2O \rightarrow 2H_3O^+ + SO_4^{2-}$
Dibasic acid, so $[H^+] = 2[H_2SO_4]$
$= 0.20$ mol dm⁻³
$pH = -\log(0.2) = 0.70$

(c) Write an equation for the second dissociation of sulfuric acid and suggest why it is far from complete.

$HSO_4^- + H_2O \rightleftharpoons H_3O^+ + SO_4^{2-}$
The H_3O^+ ions from the first dissociation suppress this equilibrium.

The assumption that a monobasic strong acid is fully dissociated is a good one, but this is not accurate for dibasic or tribasic acids. For these, some of the acid molecules will not be fully dissociated, so the $[H^+]$ above is higher than it would be and the pH is lower than it would be.

If $[H^+] = 1.0$ mol dm⁻³, then pH = 0.
If $[H^+] > 1.0$ mol dm⁻³, then pH is negative.

Now try this

1 Calculate the pH of:
 (a) 0.020 mol dm⁻³ HNO_3 **(1 mark)**
 (b) the mixture formed when 50 cm³ of 0.2 mol dm⁻³ HCl has 60 cm³ of 0.10 mol dm⁻³ NaOH added. **(3 marks)**

2 Calculate $[H^+]$ in a solution with pH = 2.4. **(1 mark)**

The ionic product of water

The ionic product of water, K_w, is the product of the concentration of H^+ and OH^- ions in water, and is used to calculate the pH of alkaline solutions.

The dissociation of water

When pure water dissociates it is **neutral** because $[H^+] = [OH^-]$

one water molecule donates a proton to another water molecule

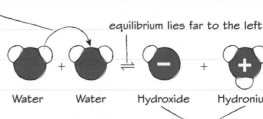

equilibrium lies far to the left

Water   Water   Hydroxide   Hydronium

equal numbers of H^+ ions and OH^- ions are formed

K_w

The equilibrium constant for the dissociation above is the ionic product of water, K_w.

$K_w = [H^+][OH^-]$

Like all equilibrium constants, it is only altered by a change of temperature.

At 25 °C:

$K_w = 1.0 \times 10^{-14}$ mol² dm⁻⁶

> 🖩 **Maths skills** **pH of pure water**
>
> In pure water, $[H^+] = [OH^-]$
>
> At 25 °C, $K_w = [H^+][OH^-] = 1.0 \times 10^{-14}$
>
> $[H^+]^2 = 1.0 \times 10^{-14}$
>
> $[H^+] = \sqrt{1.0 \times 10^{-14}}$
>
> $= 1.0 \times 10^{-7}$ mol dm⁻³
>
> pH $= -\log[H^+]$
>
> $= 7.0$

Consider, when raising the temperature, whether the position of equilibrium has moved to the left or to the right.

Worked example

pH of strong bases

Calculate the pH of 0.15 mol dm⁻³ NaOH at 25 °C.
(2 marks)

$NaOH \rightarrow Na^+ + OH^-$

$[OH^-] = [NaOH] = 0.15$ mol dm⁻³

$K_w = [H^+][OH^-] = 1.0 \times 10^{-14}$ mol² dm⁻⁶

$[H^+] = \dfrac{1.0 \times 10^{-14}}{0.15}$

     $= 6.67 \times 10^{-14}$ mol dm⁻³

pH $= -\log(6.67 \times 10^{-14}) = 13.2$

Calculate the pH of 0.10 mol dm⁻³ Ca(OH)₂ at 25 °C.
(3 marks)

$Ca(OH)_2 \rightarrow Ca^{2+} + 2OH^-$

$[OH^-] = 2[Ca(OH)_2] = 0.20$ mol dm⁻³

$K_w = [H^+][OH^-] = 1.0 \times 10^{-14}$ mol² dm⁻⁶

$[H^+] = \dfrac{1.0 \times 10^{-14}}{0.20}$

     $= 5 \times 10^{-14}$ mol dm⁻³

pH $= -\log(5 \times 10^{-14}) = 13.3$

Now try this

1   Calculate the pH of:
   (a)   0.020 mol dm⁻³ KOH.   **(1 mark)**
   (b)   0.15 mol dm⁻³ Ca(OH)₂   **(2 marks)**
   (c)   the mixture formed when 80 cm³ of 0.25 mol dm⁻³ NaOH has 60 cm³ of 0.10 mol dm⁻³ HCl added.   **(3 marks)**

2   (a)   Calculate the pH of pure water at 91 °C, when $K_w = 1.52 \times 10^{-13}$ mol² dm⁻⁶   **(2 marks)**
   (b)   Explain, using K_w at 91 °C and K_w at 25 °C = 1.0×10^{-14} mol² dm⁻⁶, whether the dissociation of water is exothermic or endothermic. **(2 marks)**
   (c)   Explain why pure water at 91 °C is neutral, when pH $\neq$ 7.   **(1 mark)**

The acid dissociation constant

Acids dissociate in water, donating protons, in an equilibrium reaction. The equilibrium constant for this reaction is the acid dissociation constant, K_a.

The dissociation equilibrium

for weak acids, equilibrium lies far to the left

conjugate base ion

The acid dissociation constant, K_a

$$HA + H_2O \rightleftharpoons H_3O^+ + A^-$$

$$K_a = \frac{[H_3O^+][A^-]}{[HA]}$$

Remember: $[H_2O]$ is omitted from K_a.

Examples:

ethanoic acid, $K_a = 1.76 \times 10^{-5}$ mol dm^{-3}

chloroethanoic acid, $K_a = 1.4 \times 10^{-3}$ mol dm^{-3}

The K_a values are at 25 °C. Like all equilibrium constants, K_a is altered by changing temperature.

> HA represents any acid. H is the hydrogen which will be released as an H^+ ion and A is the rest of the molecule.

pK_a

$pK_a = -\log[K_a]$

$K_a = 10^{-pK_a}$

Examples:

ethanoic acid, $pK_a = 4.75$

chloroethanoic acid, $pK_a = 2.85$

> • Chloroethanoic acid is a stronger acid (more dissociated) so K_a is higher, pK_a is lower than ethanoic acid.
> • The reason for this is the electron-withdrawing nature of the chlorine atom (see page 23).

Worked example

Using K_a to calculate the pH of a weak acid

Calculate the pH of 0.10 mol dm^{-3} ethanoic acid. $K_a = 1.76 \times 10^{-5}$ mol dm^{-3} **(3 marks)**

1 $CH_3COOH + H_2O \rightleftharpoons H_3O^+ + CH_3COO^-$

$$K_a = \frac{[H_3O^+][CH_3COO^-]}{[CH_3COOH]}$$

Approximation 1: $[H_3O^+] = [CH_3COO^-]$

$$= \frac{[H_3O^+]^2}{[CH_3COOH]}$$

Approximation 2: [HA] is the same as the undissociated value.

2 $[H_3O^+]^2 = K_a\,[CH_3COOH]$

$= 1.76 \times 10^{-5} \times 0.10$

$= 1.76 \times 10^{-6}$ mol^2 dm^{-6}

See page 90 for how to calculate the pH of strong acids.

3 $[H_3O^+] = \sqrt{(1.76 \times 10^{-6})}$

$= 1.33 \times 10^{-3}$ mol dm^{-3}

4 $pH = -\log(1.33 \times 10^{-3})$

$= 2.88$

> For part (b), change the pH into $[H^+]$ using the method on page 91. Then substitute this into a rearranged K_a expression. Use the same assumptions as above.

Now try this

(a) Calculate the pH of 0.010 mol dm^{-3} chloroethanoic acid. $K_a = 1.4 \times 10^{-3}$ mol dm^{-3} **(3 marks)**

(b) Calculate the concentration of chloroethanoic acid that has a pH of 3.0. **(3 marks)**

Approximations made in weak acid pH calculations

The approximations made in pH calculations for weak acids have limitations.

Worked example

Using K_a to calculate the pH of a weak acid

Calculate the pH of $0.020 \text{ mol dm}^{-3}$ propanoic acid. $K_a = 1.3 \times 10^{-5} \text{ mol dm}^{-3}$ **(3 marks)**

1 $CH_3CH_2COOH + H_2O$
$$\rightleftharpoons H_3O^+ + CH_3CH_2COO^-$$

$$K_a = \frac{[H_3O^+][CH_3CH_2COO^-]}{[CH_3CH_2COOH]}$$

$$= \frac{[H_3O^+]^2}{[CH_3CH_2COOH]}$$

2 $[H_3O^+]^2 = K_a [CH_3CH_2COOH]$
$$= 1.3 \times 10^{-5} \times 0.020$$
$$= 2.6 \times 10^{-7} \text{ mol}^2 \text{ dm}^{-6}$$

3 $[H_3O^+] = \sqrt{(2.6 \times 10^{-7})}$
$$= 5.1 \times 10^{-4} \text{ mol dm}^{-3}$$

4 $pH = -\log(5.1 \times 10^{-4})$
$$= 3.3$$

Using the pH of a weak acid to calculate K_a

When calculating K_a the same expression is used but different data is substituted.

Calculate the K_a of methanoic acid.

A $0.015 \text{ mol dm}^{-3}$ solution of the acid has a pH of 2.81. **(2 marks)**

1 $HCOOH + H_2O \rightleftharpoons H_3O^+ + HCOO^-$

$$K_a = \frac{[H_3O^+][HCOO^-]}{[HCOOH]} = \frac{[H_3O^+]^2}{[HCOOH]}$$

2 $[H_3O^+] = 10^{-pH} = 10^{-2.81}$
$$= 1.55 \times 10^{-3} \text{ mol dm}^{-3}$$

3 $K_a = \frac{(1.55 \times 10^{-3})^2}{0.015}$
$$= 1.60 \times 10^{-4} \text{ mol dm}^{-3}$$

Approximations used in the pH calculation

1 In Step 1 the approximation is made that $[CH_3CH_2COO^-] = [H_3O^+]$
- One molecule of the acid dissociates to form one of each of the ions.
- However, there will be additional H_3O^+ ions from the dissociation of water (see page 91).
- The number of additional H_3O^+ ions from the dissociation of water is negligible.

2 In Step 2, the value used for the concentration of the acid is $0.020 \text{ mol dm}^{-3}$.
- Some of the molecules of acid dissociate.
- The number is small for a weak acid.

Almost all examples of weak acids you will meet are carboxylic acids. Although they are weak, they have typical acid reactions. See page 128.

Limitations of the assumptions

The dissociation of many weak acids means that these assumptions are reasonable:
- the dissociation is sufficiently high so that most of the H^+ ions come from the acid, so that the dissociation of water is negligible
- the dissociation is sufficiently low that the concentration of acid molecules at equilibrium is near to the original value (about 2.5% of propanoic acid molecules in the first worked example).

However, for weak acids with a higher K_a value (one where more molecules dissociate) the second approximation may not be reasonable, and a more complex calculation must be used.

Which molecules release H^+ ions most easily and why? See the carboxylic acids on page 128.

Now try this

1 Calculate the K_a of an acid if a $0.010 \text{ mol dm}^{-3}$ solution of the acid has a pH of 2.93. **(3 marks)**
2 Calculate the pH of 0.5 mol dm^{-3} butanoic acid. $pK_a = 4.82$ **(3 marks)**

Buffers

A buffer solution is a mixture that minimises pH changes on the addition of small amounts of acid or base.

Buffer composition

- A buffer solution may be made from a weak acid and its salt.
- An example is a mixture of ethanoic acid and sodium ethanoate.
- Only a small proportion of the ethanoic acid molecules are dissociated.

$$CH_3COOH + H_2O \rightleftharpoons CH_3COO^- + H_3O^+$$

- The salt is completely dissociated.

$$CH_3COONa \rightarrow CH_3COO^- + Na^-$$

- Different mixtures of acid and salt give buffers of different pH value – see page 95.

1. The acid in a buffer solution is always weak, so it will only be slightly dissociated.

2. Adding the salt adds more of the conjugate base ions (in this example ethanoate ions) which means that the position of equilibrium moves left (the dissociation is **suppressed**) so even fewer acid molecules are dissociated.

3. So, in the mixture there will be a large amount (**reservoir**) of undissociated acid molecules (from the acid) and conjugate base ions (from the salt).

Buffer action when acid added

1. The added H^+ ions react with conjugate base ions:

$$CH_3COO^- + H_3O^+ \rightarrow CH_3COOH + H_2O$$

2. As there is a large reservoir of conjugate base ions, almost all of the H^+ ions are removed, leaving the pH little changed.

Buffer action when base added

1. The added OH^- ions react with acid molecules:

$$CH_3COOH + OH^- \rightarrow CH_3COO^- + H_2O$$

2. As there is a large reservoir of acid molecules, almost all of the OH^- ions are removed, leaving the pH little changed.

Making a buffer solution

There are two ways of making a buffer solution:

1. Take a weak acid, and add a measured amount of the solid salt. For example, dissolve some sodium ethanoate crystals in some dilute ethanoic acid.

2. Take a weak acid and partly neutralise it with a strong alkali. For example, add sodium hydroxide solution to dilute ethanoic acid.

$$CH_3COOH + NaOH \rightarrow CH_3COONa + H_2O$$

An excess of the acid is used, so the mixture after the part-neutralisation contains the unreacted weak acid and the salt that has been formed.

Buffering in the blood

acidosis or alkalosis

0 .... 7 14

death   7.35–7.45 normal   death
               pH range of blood

- The blood plasma contains a buffer consisting of the weak acid, carbonic acid, H_2CO_3, and its conjugate base, hydrogencarbonate ions, HCO_3^-.
- Carbonic acid (weak acid): $H_2CO_3 + H_2O \rightleftharpoons HCO_3^- + H_3O^+$
- During exercise the body produces lactic acid, increasing the blood's $[H^+]$, eating acidic or alkaline foods also alters blood pH.
- If H^+ ions are added to the blood: $HCO_3^- + H^+ \rightarrow H_2CO_3$.
- If OH^- ions are added to the blood: $H_2CO_3 + OH^- \rightarrow HCO_3^- + H_2O$.

Worked example

Which mixture would make a buffer?   **(1 mark)**

A hydrochloric acid and sodium chloride
B propanoic acid and sodium chloride
C propanoic acid and sodium propanoate
D hydrochloric acid and sodium propanoate

C

Now try this

Explain how the blood maintains an almost constant pH if vigorous exercise is undertaken.   **(3 marks)**

Buffer calculations

Different mixtures of a weak acid and its salt will make buffer solutions with a pH that can be calculated.

Maths skills Calculating the pH of buffers

The acid dissociation constant for the weak acid, HA, applies even when it is mixed with other substances.

$$K_a = \frac{[H_3O^+][A^-]}{[HA]}$$

In a buffer, if $[A^-] = [HA]$, then these terms cancel and $K_a = [H_3O^+]$, so $pK_a = pH$

• When the sodium hydroxide neutralises some of the acid (in a 1:1 ratio) it is all used up, so the moles of acid neutralised = original moles of alkali. The moles of salt formed (1:1 ratio) = moles of acid neutralised.

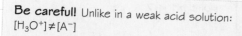

Be careful! Unlike in a weak acid solution: $[H_3O^+] \neq [A^-]$

Worked example

Making a buffer solution by partly neutralising an acid

Calculate the pH of the buffer solution made by mixing 50.0 cm³ of 0.220 mol dm⁻³ butanoic acid with 0.20 g of pure sodium hydroxide.   **(4 marks)**
$K_a = 1.5 \times 10^{-5}$ mol dm⁻³

1 Initial mol acid = $50.0 \times \frac{0.220}{1000} = 0.0110$ mol

2 Initial mol alkali = $\frac{0.20}{40} = 0.0050$ mol

3 After neutralisation,
mol acid = $0.0110 - 0.0050 = 0.0060$ mol

4 After neutralisation, mol salt = 0.0050 mol

5 $K_a = \frac{[H_3O^+][\text{sodium butanoate}]}{[\text{butanoic acid}]}$

$1.5 \times 10^{-5} = \frac{[H_3O^+]\left(\frac{0.0050}{0.050}\right)}{\left(\frac{0.0060}{0.050}\right)}$

6 $[H_3O^+] = \dfrac{1.5 \times 10^{-5} \times \left(\frac{0.0060}{0.050}\right)}{\left(\frac{0.0050}{0.050}\right)}$

$= 1.8 \times 10^{-5}$

7 $pH = -\log[H^+] = -\log(1.8 \times 10^{-5}) = 4.74$

Worked example

Calculate the pH of the buffer solution formed when 50 cm³ of 0.10 mol dm⁻³ ethanoic acid is mixed with 150 cm³ of 0.20 mol dm⁻³ sodium ethanoate solution. $K_a = 1.76 \times 10^{-5}$ mol dm⁻³   **(4 marks)**

1 In mixture,   The total volume is 200 cm³
$[CH_3COOH] = 0.1 \times \frac{50}{200} = 0.025$ mol dm⁻³

2 $[CH_3COONa] = 0.20 \times \frac{150}{200} = 0.15$ mol dm⁻³

3 $K_a = \frac{[H_3O^+][A^-]}{[HA]}$ $1.76 \times 10^{-5} = \frac{[H_3O^+](0.15)}{(0.025)}$

4 $[H_3O^+] = \frac{1.76 \times 10^{-5} \times 0.025}{0.15}$

$= 2.93 \times 10^{-6}$ mol dm⁻³

5 $pH = -\log[H^+] = -\log(2.93 \times 10^{-6}) = 5.53$

Now try this

Calculate the pH of the buffer made by mixing 100 cm³ of 0.25 mol dm⁻³ butanoic acid with 50.0 cm³ of 0.15 mol dm⁻³ NaOH.
$K_a = 1.5 \times 10^{-5}$ mol dm⁻³
                     (4 marks)

pH titration curves

The pH values when an acid is being neutralised by a base can be measured with a pH meter. The shape of the curve when the pH values are plotted is characteristic, depending on whether the acid is strong or weak and whether the base is strong or weak.

pH curve for a strong acid–strong base titration

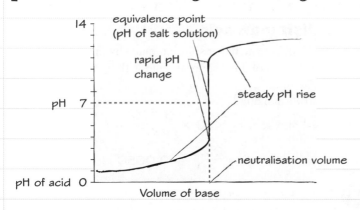

For strong acid–strong base, the pH curve is vertical for at least three pH units below 7 and three pH units above 7.

pH curves for other combinations of acid and base

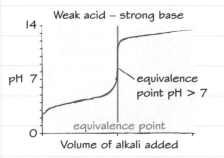

For weak acid–strong base, the pH curve is vertical for at least 3 pH units above 7.

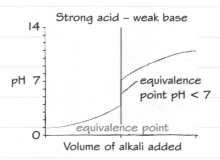

For strong acid–weak base, the pH curve is vertical for at least three pH units below 7.

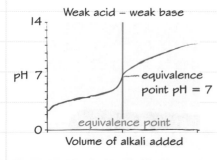

For weak acid–weak base, the pH curve is never vertical.

Worked example

Compare differences in features found on strong acid-strong base and a weak acid–strong base pH titration curves. **(3 marks)**

The starting pH is likely to be lower for a strong acid. The weak acid has a lip at the start. The weak acid gradually increases to reach 7, but the strong acid remains steady and sharply turns to pH 7, and has a vertical region from about 4–7.

Now try this

Sketch as accurately as possible the pH titration curve when 25 cm³ of 0.1 mol dm⁻³ HCl is titrated with 50 cm³ of 0.1 mol dm⁻³ NaOH. **(5 marks)**

The starting and end pH values can be calculated from the data given.

Be careful! Although this is a sketch, draw this carefully considering: start and end pH; shape; vertical regions; neutralisation volume.

Indicators

Indicators are weak acids where the acid molecule has one colour and the conjugate base ion has another colour.

How indicators work

Indicators, HIn, are weak acids.

$$\text{colour 1} \overset{\text{HIn} + H_2O \rightleftharpoons H_3O^+ + \text{In}^-}{} \text{colour 2}$$

- When added to acidic solutions, $[H_3O^+]$ is increased, the position of equilibrium moves left, and more HIn forms: colour 1.
- When added to alkaline solutions, $[H_3O^+]$ is reduced, the position of equilibrium moves right, and more In^- forms: colour 2.

The pH range of an indicator is the range of values over which the eye can detect that the indicator is changing colour.

methyl orange: 3.1–4.4

phenolphthalein: 8.3–10

> **Practical skills** **Using a pH meter**
>
> pH is measured using a pH meter. The meter must be calibrated before use.
>
> **1** Rinse the electrode with distilled water.
>
> **2** Place the electrode in a buffer solution of known pH.
>
> **3** Adjust the pH meter until it reads the correct value.
>
> **4** Rinse the electrode again, then repeat Steps 2 and 3 with a buffer solution of different pH.
>
> **5** The pH meter is now ready to use. Place the electrode into the solution, stir and allow the pH value to settle.

Choosing an indicator

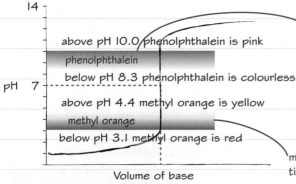

above pH 10.0 phenolphthalein is pink

phenolphthalein

below pH 8.3 phenolphthalein is colourless

above pH 4.4 methyl orange is yellow

methyl orange

below pH 3.1 methyl orange is red

Volume of base

phenolphthalein can be used in titration involving a strong base

The pH range of indicators will be given if needed in the exam paper.

methyl orange can be used in titration involving a strong acid

The indicator must show a sharp colour change. The pH range of the indicator must be within the vertical part of the pH titration curve. A weak acid–weak base pH titration curve has no vertical region (see previous page) so no indicator is suitable – they would all give a gradual colour change.

Worked example

Consider the pH titration curve.

(a) Explain whether the acid and base used were strong or weak. **(2 marks)**

The acid was weak as there is no vertical region below pH 7, and the base strong as there is a vertical region above pH 7.

(b) Suggest a suitable indicator for this titration. Explain your choice. **(2 marks)**

Phenolphthalein, because its pH range is within the vertical region of the graph so would give a sharp colour change.

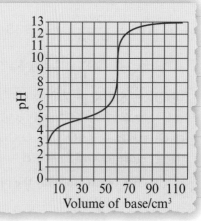

Now try this

An indicator has a pH range of 4.1–6.2.

(a) Suggest an acid–base combination for which the indicator is suitable. **(1 mark)**

(b) Explain why this indicator is **not** suitable for finding the end point of the titration between ethanoic acid and sodium hydroxide solution. **(2 marks)**

(c) Explain why there is no suitable indicator for a weak acid–weak base titration, and suggest another method of finding the end point of this titration. **(3 marks)**

Exam skills 8

This exam-style question uses knowledge and skills you have already revised. Look at pages 86, 93 and 95 for a reminder about calculating K_c, weak acid pH calculations and their limitations and buffer solutions.

Worked example

Hydrogen reacts with the element X_2 to form HX.

$H_2 + X_2 \rightleftharpoons 2HX$

3.0 mol of hydrogen and 2.0 mol of the element X_2 are mixed in a 4.0 dm³ container. At equilibrium, 1.8 mol of hydrogen remain.

> **Maths skills** Carefully practise working out the mol at equilibrium. The equation tells you the ratio of the mol that **react** (in this case 1.2:1.2:2.4), **not** the mol at equilibrium.

(a) (i) Calculate the moles of X_2 and of HX at equilibrium. **(2 marks)**

mol of hydrogen reacted = 3.0 − 1.8 = 1.2 mol

mol of X_2 remaining at equilibrium = 2.0 − 1.2 = 0.8 mol

mol of HX formed at equilibrium = 2 × 1.2 = 2.4 mol

(ii) Write an expression for the equilibrium constant, K_c. **(1 mark)**

$K_c = \dfrac{[HX]^2}{[H_2][X_2]}$

(iii) Calculate K_c.

> **Maths skills** Remember to divide by the volume. (In this case the volume cancels but it doesn't always.)

(2 marks)

$K_c = \dfrac{\left(\dfrac{2.4}{4.0}\right)^2}{\left(\dfrac{1.8}{4.0}\right)\left(\dfrac{0.8}{4.0}\right)} = \dfrac{0.36}{0.09} = 4.0$

(b) 4.0 dm³ of the gas HX is dissolved in water to make 500 cm³ of the weak acid, HX(aq). K_a for this acid is 7.2×10^{-4} mol dm⁻³.

(i) Calculate the pH of the acid. (1 mol of any gas under the conditions of this experiment has a volume 24 dm³.) **(5 marks)**

mol of HX = $\dfrac{4.0}{24}$ = 0.167 mol

$[HX] = \dfrac{0.167}{0.500} = 0.333$ mol dm⁻³

> **Maths skills** All of the calculations around weak acids (including buffer calculations) start with the expression for K_a. Write this in your answer, and then substitute values as required.

$K_a = \dfrac{[H_3O^+][X^-]}{[HX]}$ $K_a = \dfrac{[H_3O^+]^2}{[HX]}$

> $[H_3O^+] = [X^-]$ in the pure acid solution.

$[H_3O]^2 = K_a[HX] = 7.2 \times 10^{-4} \times 0.33$

$\quad = 2.4 \times 10^{-4}$ mol² dm⁻⁶

$[H_3O] = \sqrt{(2.4 \times 10^{-4})} = 0.0155$ mol dm⁻³

pH = −log (0.0155) = 1.8

(ii) Give one reason why the pH value that you have calculated is only an approximation. **(2 marks)**

The [HX] used in the calculation is 0.33 mol dm⁻³. The actual [HX] will be less than 0.33 mol dm⁻³ because HX partly dissociates in water. It is assumed that because HX is a weak acid, only a small proportion of HX dissociates and this amount can be ignored.

> **Maths skills** When writing down your answer, round the numbers suitably – for example, to one significant figure more than your final answer – but keep the whole answer in your calculator, and use the [ANS] button. For example, mol
>
>
> $[HX] = 0.166666...$ $[HX] = \dfrac{\text{ANS}}{0.5}$

(iii) Starting with a solution of HX, how could you make an acid buffer? **(2 marks)**

Add some sodium hydroxide solution to a sample of the acid so that it is partly neutralised, giving the weak acid and its salt.

The Born–Haber cycle

The **lattice enthalpy**, $\Delta_{LE}H$, is the enthalpy change when one mole of ionic lattice is formed from its gaseous ions. This cannot be measured directly, so a Hess' law cycle is used (see page 43) called the **Born–Haber** cycle.

The Born Haber cycle for sodium chloride

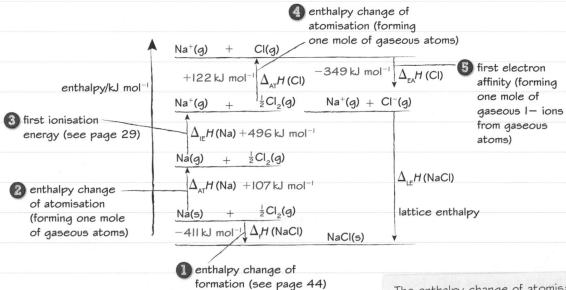

4 enthalpy change of atomisation (forming one mole of gaseous atoms)

5 first electron affinity (forming one mole of gaseous I− ions from gaseous atoms)

3 first ionisation energy (see page 29)

2 enthalpy change of atomisation (forming one mole of gaseous atoms)

1 enthalpy change of formation (see page 44)

enthalpy/kJ mol⁻¹

$Na^+(g)$ + $Cl(g)$

$+122\ kJ\ mol^{-1}$ $\Delta_{AT}H\,(Cl)$ $-349\ kJ\ mol^{-1}$ $\Delta_{EA}H\,(Cl)$

$Na^+(g)$ + $\frac{1}{2}Cl_2(g)$ $Na^+(g)\ +\ Cl^-(g)$

$\Delta_{IE}H\,(Na)\ +496\ kJ\ mol^{-1}$

$Na(g)$ + $\frac{1}{2}Cl_2(g)$

$\Delta_{AT}H\,(Na)\ +107\ kJ\ mol^{-1}$ $\Delta_{LE}H\,(NaCl)$

$Na(s)$ + $\frac{1}{2}Cl_2(g)$ lattice enthalpy

$-411\ kJ\ mol^{-1}$ $\Delta_fH\,(NaCl)$ $NaCl(s)$

Lattice enthalpy = − 5 − 4 − 3 − 2 + 1

All of the steps in this cycle can be measured by experiment except the lattice enthalpy.

Lattice enthalpies are always exothermic due to the strong electrostatic attraction between oppositely charged ions.

> The enthalpy change of atomisation is the enthalpy change when one mole of gaseous atoms is **formed**, not when one mole of substance is atomised. For example, for chlorine the equation is $\frac{1}{2}Cl_2(g) \rightarrow Cl(g)$

> You must know the definitions of first ionisation energy (see page 29) and enthalpy change of formation (see page 46).

Worked example

Calculate the lattice enthalpy of sodium chloride using the data given in the cycle above. **(2 marks)**

Lattice enthalpy
= − 5 − 4 − 3 − 2 + 1
= 349 − 122 − 496 − 107 − 411
= −787 kJ mol⁻¹

Now try this

(a) Draw a Born–Haber cycle for the formation of potassium chloride. **(2 marks)**

> Check that every level has the same number of K and Cl particles.

(b) Calculate the lattice enthalpy of potassium chloride using the data in the cycle above, and the following data: **(2 marks)**

enthalpy change of atomisation (K) +77 kJ mol⁻¹
1st ionisation energy (K) +419 kJ mol⁻¹
enthalpy change of formation (KCl) −437 kJ mol⁻¹

Factors affecting lattice enthalpy

The lattice enthalpy is a measure of the strength of the ionic bonding in a giant ionic lattice (see page 19).

The Born–Haber cycle for magnesium chloride is shown.

(a) What enthalpy change is represented by ΔH_1?
(1 mark)

The second ionisation energy of magnesium.

(b) What enthalpy change is represented by ΔH_2?
(1 mark)

$2 \times$ the enthalpy change of atomisation of chlorine.

This is also the bond enthalpy of Cl_2 (see page 46).

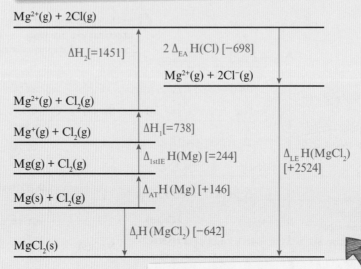

Note that the **second** electron has been removed.

$Mg^{2+}(g) + 2Cl(g)$

$\Delta H_2 [=1451]$ $2\,\Delta_{EA} H(Cl)\,[-698]$

$Mg^{2+}(g) + 2Cl^-(g)$

$Mg^{2+}(g) + Cl_2(g)$

$Mg^+(g) + Cl_2(g)$ $\Delta H_1 [=738]$

$Mg(g) + Cl_2(g)$ $\Delta_{1stIE} H(Mg) [=244]$ $\Delta_{LE} H(MgCl_2)\ [+2524]$

$Mg(s) + Cl_2(g)$ $\Delta_{AT} H(Mg) [+146]$

$\Delta_f H(MgCl_2) [-642]$

$MgCl_2(s)$

Take special care with the multiples: two moles of Cl atoms form ions.

(c) Use the cycle and the data provided to calculate the electron affinity of chlorine, $\Delta_{EA}H(Cl)$.   **(2 marks)**

$2 \times \Delta_{EA}H(Cl) = -1451 - 738 - 244 - 146 - 642 + 2524 = -697$

$\Delta_{EA}H(Cl) = \dfrac{-697}{2} = -348.5\ kJ\ mol^{-1}$

Ion size

Larger ions will be further apart in the lattice, so reducing electrostatic attraction (weakening ionic bonding), giving a less exothermic lattice enthalpy.

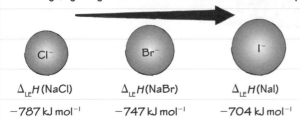

$\Delta_{LE}H(NaCl)$	$\Delta_{LE}H(NaBr)$	$\Delta_{LE}H(NaI)$
$-787\ kJ\ mol^{-1}$	$-747\ kJ\ mol^{-1}$	$-704\ kJ\ mol^{-1}$

Ion charge

More highly charged ions will have increased electrostatic attraction (strengthening ionic bonding), giving a more exothermic lattice enthalpy.

	OH^-	O^{2-}	increasing charge
Na^+	-900	-2481	↓
Mg^{2+}	-3006	-3791	
Al^{3+}	-5627	-15916	
increasing charge	⟶		

A comparison of NaCl with MgCl₂

- Magnesium ions have a higher charge than sodium ions.
- Magnesium ions are smaller than sodium ions.
- There is a greater electrostatic attraction between the ions in $MgCl_2$.
- The lattice enthalpy for $MgCl_2$ is more exothermic.

$\Delta_{LE}H(NaCl) = -787\ kJ\ mol^{-1}$

$\Delta_{LE}H(MgCl_2) = -2524\ kJ\ mol^{-1}$

Arrange the following substances in order of increasingly exothermic lattice enthalpy. Explain the reason for the order.   **(4 marks)**

lithium chloride, lithium oxide, potassium chloride, potassium bromide, potassium iodide, sodium chloride

The enthalpy change of solution

Some ionic solids dissolve in water and some do not. A Hess' law cycle (see page 43) can be drawn to investigate the enthalpy change when a solid dissolves.

Maths skills **Hess' law cycle for dissolving an ionic solid, MX**

$$\Delta_{sol}H = -\Delta_{LE}H + \Sigma\Delta_{hyd}H$$

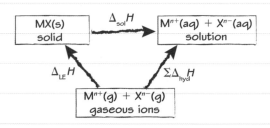

- The enthalpy change of solution, $\Delta_{sol}H$, is the enthalpy change when 1 mol of solid is dissolved to make an infinitely dilute solution.
- The enthalpy change of hydration $\Delta_{hyd}H$, is the enthalpy change when 1 mol of gaseous ions is dissolved to make an infinitely dilute solution.

The definition for lattice enthalpy is on page 99.

Dissolving a solid in water causes an enthalpy change. An infinitely dilute solution is one where adding more water to the solution gives no further enthalpy change. You do not need to know this definition in the exam.

Enthalpy level diagram for dissolving

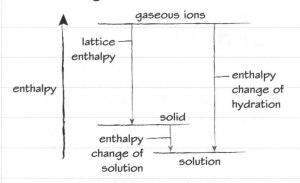

This enthalpy level diagram is for an exothermic dissolving reaction.

Typically, the more exothermic the enthalpy change of solution, the more soluble a solid is.

Factors affecting enthalpy change of hydration

1 Ion charge
More highly charged ions will have increased electrostatic attraction to water molecules, giving a more exothermic enthalpy change of hydration.

2 Ion size
Larger ions have a reduced electrostatic attraction to water molecules, giving a less exothermic enthalpy change of hydration.

Factors affecting $_{hyd}H$ **and** $_{LE}H$

More highly charged ions → more exothermic $\Delta_{hyd}H$ and $\Delta_{LE}H$

Larger ions → less exothermic $\Delta_{hyd}H$ and $\Delta_{LE}H$

(See page 100 for a reminder about this topic.)

Worked example

Calculate $\Delta_{sol}H$ for NaCl given the following data. **(2 marks)**

lattice enthalpy (NaCl) -781 kJ mol^{-1}

enthalpy change of hydration (Na$^+$) -406 kJ mol^{-1}

enthalpy change of hydration (Cl$^-$) -378 kJ mol^{-1}

$\Delta_{sol}H = -\Delta_{LE}H + \Sigma\Delta_{hyd}H$
$= +787 - 406 - 378$
$= +3$ kJ mol^{-1}

Now try this

Calculate $\Delta_{sol}H$ for MgCl$_2$ given the following data. **(2 marks)**

lattice enthalpy (MgCl$_2$) -2524 kJ mol^{-1}
enthalpy change of hydration (Mg^{2+}) -1926 kJ mol^{-1}
enthalpy change of hydration (Cl$^-$) -378 kJ mol^{-1}

Remember that there are 2 mol of chloride ions.

Entropy

Entropy (S) is a measure of the dispersal of the energy in a system. The more disordered the system, the greater the entropy.

Entropy in different states

$H_2O(s)$	$H_2O(l)$	$H_2O(g)$
45 J K^{-1} mol^{-1}	70 J K^{-1} mol^{-1}	189 J K^{-1} mol^{-1}

→ increasing entropy

Entropy is measured in J K^{-1} mol^{-1}

1 molecules in ordered solid lattice – low entropy

2 molecules disordered in liquid – higher entropy

3 molecules very disordered in gas – highest entropy

Entropy values

- Entropy of a system depends on the temperature – the higher the temperature the higher the entropy.
- In general, gases have a higher entropy than liquids which, have a higher entropy than solids.

Calculating changes in entropy of the system

- The entropy change during a reaction:
  $$\Delta S = S(\text{products}) - S(\text{reactants})$$
- If there is an increase in the number of moles of gas, there will usually be an increase in entropy.

Worked example

Butan-1-ol can be fully combusted to form carbon dioxide and water.

(a) Write the balanced equation for the reaction.
(2 marks)

$$CH_3(CH_2)_3OH(l) + 6O_2(g) \rightarrow 4CO_2(g) + 5H_2O(l)$$

(b) Calculate the entropy change of the system for the reaction using this data: **(3 marks)**

	S/J K^{-1} mol^{-1}
$CH_3(CH_2)_3OH(l)$	228
$O_2(g)$	205
$CO_2(g)$	214
$H_2O(l)$	70

$\Delta S = S(\text{products}) - S(\text{reactants})$
$= ((4 \times 214) + (5 \times 70))$
$\quad - (228 + (6 \times 205))$
$= 1206 - 1458$
$= -252$ J K mol^{-1}

(c) Explain whether your answer has the expected sign. **(1 mark)**

There are 6 moles of gaseous reactant and 4 moles of gaseous product, so with a reduction in moles of gas there is expected to be a reduction in entropy, so the sign is negative as expected.

Now try this

1 Use the data on the left to calculate the entropy change of the system for the reaction in which butan-1-ol burns in a limited supply of air to produce carbon monoxide and water only.
(3 marks)

$S(CO) = 198$ J K^{-1} mol^{-1}

Write the balanced equation first.

2 Place the following in order of increasing entropy, and explain your answer. **(3 marks)**
The carbon dioxide and water vapour produced by burning 1 g of wax
1 g of liquid wax at 40°C
1 g of liquid wax at 60°C
1 g of solid wax at 40°C

Lay out your answer clearly showing entropy of reactant and products separately. This will make it easier to work out your answer.

In questions like this, consider whether there has been a change in the number of moles of gas.

Free energy

A process is *feasible* when the free energy change, ΔG, has a negative value.

The Gibbs equation

The free energy change for a process, ΔG, depends on:

- the entropy change, ΔS (see page 102)
- the temperature, T, in K
- the enthalpy change, ΔH (see page 40).

These are combined in the Gibbs equation:

$\Delta G = \Delta H - T\Delta S$

Reaction feasibility

The feasibility of a process depends on the value of ΔG.

- ΔG positive: process is not feasible
- $\Delta G = 0$: process can form an equilibrium
- ΔG negative: process is feasible.

Feasible means that the process can occur. However, many feasible processes have a high activation energy and will be very slow or even so slow as to be unobservable.

Feasibility and temperature

$4Ag(s) + O_2(g) \rightarrow 2Ag_2O(s)$

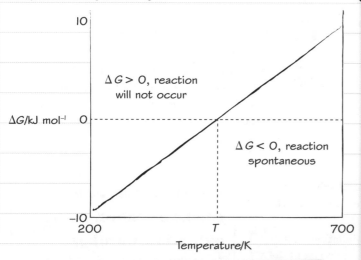

$\Delta G/kJ\,mol^{-1}$

$\Delta G > 0$, reaction will not occur

$\Delta G < 0$, reaction spontaneous

Temperature/K

Above temperature T the reaction will not occur. Here, the reverse reaction is feasible.

Be careful! S values are in $J\,K^{-1}\,mol^{-1}$, so divide by 1000 to convert into $kJ\,K^{-1}\,mol^{-1}$ before substituting into the Gibbs equation.

Worked example

🖩 Maths skills   **Calculation of the free energy change, ΔG**

Calculate the temperature, T, below which the reaction $4Ag(s) + O_2(g) \rightarrow 2Ag_2O(s)$ is feasible.

$\Delta H = -62.2\,kJ\,mol^{-1}$
$S(Ag) = 42.6\,J\,K^{-1}\,mol^{-1}$
$S(O_2) = 205.2\,J\,K^{-1}\,mol^{-1}$
$S(Ag_2O) = 121.3\,J\,K^{-1}\,mol^{-1}$

$\Delta S = (2 \times 121.3) - ((4 \times 42.6) + 205.2)$
$\quad = -133\,J\,K^{-1}\,mol^{-1}$
$\quad = -0.133\,kJ\,K^{-1}\,mol^{-1}$

$\Delta G = \Delta H - T\Delta S$
$0 = -62.2 - T(-0.133)$
$T(-0.133) = -62.2$
$T = \dfrac{-62.2}{-0.133}$
$\quad = 468\,K$

Now try this

Calcium and magnesium are normally extracted by electrolysis but it is feasible that calcium oxide could be reduced by carbon: $CaO(s) + C(s) \rightarrow Ca(s) + CO(g)$

Use the data in the table below to help you answer parts (a)–(c).

	CaO(s)	C(s)	Ca(s)	CO(g)
$\Delta H_f^{\ominus}$ /kJ mol^{-1}	−635	0	0	−110
$S^{\ominus}$/J K^{-1} mol^{-1}	39.7	5.7	41.4	197.6

(a) Calculate the standard enthalpy change for the CaO reduction in the equation.   **(2 marks)**

(b) Calculate the standard entropy change for the CaO reduction in the equation.   **(2 marks)**

(c) Calculate the minimum temperature at which the reduction in the equation is feasible.   **(3 marks)**

Redox

A reaction in which there is a transfer of electrons is called a redox reaction (see page 15) and it involves an **oxidising agent** and a **reducing agent**.

> OILRIG: Oxidation Is Loss, Reduction Is Gain (of electrons)

A redox reaction

Zinc will displace copper from a solution of copper(II) sulfate.

each zinc atom loses 2 electrons; the zinc atom is oxidised — oxidation

each copper ion gains 2 electrons; the copper ion is reduced

$$Zn(s) + Cu^{2+}(aq) \rightleftharpoons Zn^{2+}(aq) + Cu(s)$$

reduction

 The sulfate ions remain in solution and take no part in the reaction – they are **spectator ions**.

Oxidising and reducing agents

1 An oxidising agent is a species that accepts electrons (given by the species being oxidised). In the example Cu^{2+} is the oxidising agent.

2 A reducing agent is a species that donates electrons (to the species being reduced). In the example, Zn is the reducing agent.

> The oxidising agent is itself reduced.
> The reducing agent is itself oxidised.

Half-equations

Half-equations show either the oxidation or the reduction.

1 Oxidation: electrons are produced.
$Zn \rightarrow Zn^{2+} + 2e^-$

2 Reduction: electrons are used.
$Cu^{2+} + 2e^- \rightarrow Cu$

3 When half-equations are combined to give the equation for the redox reaction, there must be the same number of electrons in each half-equation (by multiplying as necessary) before the half-equations are combined.

Worked example

1 Deduce the ionic equation of the reaction when peroxodisulfate ions react with manganese(II) ions. The half-equations are given. **(2 marks)**
$S_2O_8^{2-} + 2e^- \rightarrow 2SO_4^{2-}$
$Mn^{2+} + 4H_2O \rightarrow MnO_4^- + 8H^+ + 5e^-$

1 Peroxodisulfate half-equation × 5
$5S_2O_8^{2-} + 10e^- \rightarrow 10SO_4^{2-}$

> Half-equations will help you to work out the ionic equation of a reaction.

2 Manganese half-equation × 2
$2Mn^{2+} + 8H_2O \rightarrow 2MnO_4^- + 16H^+ + 10e^-$

3 Add equations and cancel equal electrons
$5S_2O_8^{2-} + 2Mn^{2+} + 8H_2O$
$\rightarrow 2MnO_4^- + 16H^+ + 10SO_4^{2-}$

2 Deduce the ionic equation of the reaction when manganate ions react with iron(II) ions. The half-equations are given. **(2 marks)**
$MnO_4^- + 8H^+ + 5e^- \rightarrow Mn^{2+} + 4H_2O$
$Fe^{2+} \rightarrow Fe^{3+} + e^-$

1 Oxidation number of iron:
$Fe^{2+} = 2 \rightarrow Fe^{3+} = 3$; increase of 1.

2 Oxidation number of manganese:
$MnO_4^- = 7 \rightarrow Mn^{2+} = 2$; reduction of 5.

3 Use 5 × iron half-equation to match charges.
$MnO_4^- + 8H^+ + 5e^- + 5Fe^{2+}$
$\rightarrow Mn^{2+} + 4H_2O + 5Fe^{3+} + 5e^-$

4 Cancel electrons on each side.
$MnO_4^- + 8H^+ + 5Fe^{2+} \rightarrow Mn^{2+} + 4H_2O + 5Fe^{3+}$

Construction of a redox equation using oxidation numbers

The total increase in oxidation number of the oxidised species = the total reduction in oxidation number of the reduced species

Now try this

Deduce the ionic equation of the reaction when peroxodisulfate ions react with iron(II) ions. The half-equations are given. **(2 marks)**
$S_2O_8^{2-} + 2e^- \rightarrow 2SO_4^{2-}$
$Fe^{2+} \rightarrow Fe^{3+} + e^-$

 See page 14 for calculation of oxidation number.

Redox titrations

Titration can be used to determine the concentration of a solution in an acid–base reaction (see page 11) and also in a redox reaction. Two examples of redox reactions are shown.

 Practical skills **Manganate/ iron(II) titrations**

1 Rinse a pipette with a solution containing Fe^{2+} ions then dispense $25cm^3$ of the solution into a conical flask.

2 Rinse and fill the burette with potassium manganate(VII) solution.

> The manganate(VII) solution is deep purple – it is tricky to judge the position of the **meniscus**.

3 Add the manganate(VII) solution, swirling the flask, a drop at a time towards the end-point, until a permanent pink colour is seen.

> The colour of the excess manganate(VII) can be seen so no indicator is needed.

4 Repeat the procedure to achieve **concordant** results.

 Practical skills **Iodine/ thiosulfate titrations**

1 Rinse a pipette with a solution containing iodine then dispense $25cm^3$ of the solution into a conical flask.

2 Rinse and fill the burette with thiosulfate solution.

3 Add the thiosulfate solution, swirling the flask, a drop at a time towards the end-point.

4 Towards the end-point, add a few drops of starch solution. Add thiosulfate solution until the blue-black colour disappears.

> When dilute iodine solution is pale yellow and hard to see, so starch is added to give a darker colour.

5 Repeat the procedure to achieve concordant results.

Worked example

Titration calculation

1 Five grams of impure $FeSO_4 \cdot 7H_2O$ is made into a 250 cm^3 solution; 25.0 cm^3 of this solution is titrated with $0.0180 \text{ mol dm}^{-3}$ potassium manganate(VII) solution. The titre value is 19.6 cm^3. Calculate the percentage purity of the sample. **(6 marks)**

1 The half-equations are:
$MnO_4^- + 8H^+ + 5e^- \rightarrow Mn^{2+} + 4H_2O$
$Fe^{2+} \rightarrow Fe^{3+} + e^-$
Ratio is $1\ MnO_4^- : 5\ Fe^{2+}$ (see page 104).

2 mol of MnO_4^- = $19.6/1000 \times 0.0180$
= 3.528×10^{-4} mol

3 mol of Fe^{2+} (in 25.0 cm^3 of solution)
= $5 \times 3.528 \times 10^{-4} = 1.764 \times 10^{-3}$ mol

4 mol of Fe^{2+} (in 250 cm^3 of solution)
= $10 \times 1.764 \times 10^{-3} = 1.764 \times 10^{-2}$ mol

5 mass of $FeSO_4 \cdot 7H_2O = 277.9 \times 1.764 \times 10^{-2}$
= 4.904 g

6 percentage purity = $4.904/5.00 \times 100\%$
= 98.1%

2 25.0 cm^3 of a solution containing iodine at $0.500 \text{ mol dm}^{-3}$ is titrated with sodium thiosulfate solution; 28.6 cm^3 is needed. Calculate the concentration of the sodium thiosulfate solution. **(3 marks)**

1 The half-equations are:
$I_2 + 2e^- \rightarrow 2I^-$
$2S_2O_3^{2-} \rightarrow S_4O_6^{2-} + 2e^-$
Ratio is $1\ I_2 : 2\ S_2O_3^{2-}$

2 mol of iodine
= $25.0/1000 \times 0.500$
= 0.0125 mol

3 mol of $S_2O_3^{2-} = 2 \times 0.0125$
= 0.025 mol

4 concentration of $S_2O_3^{2-}$
= $0.025/0.0288$
= $0.874 \text{ mol dm}^{-3}$

Now try this

The VO_2^+ ion is reduced to V^{n+} by zinc and dilute sulfuric acid.
25.0 cm^3 of $0.100 \text{ mol dm}^{-3}$ solution containing VO_2^+ is reduced to V^{n+}. The resulting solution is titrated with $0.0500 \text{ mol dm}^{-3}$ MnO_4^- and 30.0 cm^3 is required to oxidise the V^{n+} back to VO_2^+. Find n. **(5 marks)**

Electrochemical cells

An electrochemical cell is a cell in which the oxidation reduction of a redox reaction are separated, and the transfer of electrons forms a current.

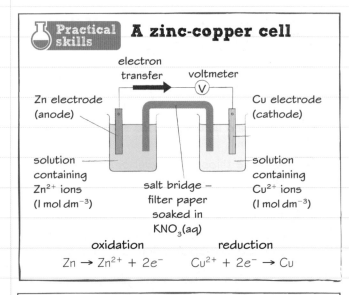

Practical skills

A zinc-copper cell

electron transfer   voltmeter (V)

Zn electrode (anode)

Cu electrode (cathode)

solution containing Zn^{2+} ions ($1 \, mol \, dm^{-3}$)

solution containing Cu^{2+} ions ($1 \, mol \, dm^{-3}$)

salt bridge – filter paper soaked in $KNO_3(aq)$

oxidation   reduction
$Zn \rightarrow Zn^{2+} + 2e^-$ $Cu^{2+} + 2e^- \rightarrow Cu$

- Metals react to form positive ions.
- In the cell above, zinc or copper could possibly lose electrons.

  $Zn \rightarrow Zn^{2+} + 2e^-$ **or** $Cu \rightarrow Cu^{2+} + 2e^-$

- When tested, Zn loses electrons, and so copper(II) ions must gain electrons,

  $Cu^{2+} + 2e^- \rightarrow Cu$. (Zinc is above copper in the reactivity series.)

- Electrons flow round the circuit from left to right.
- Ions in the salt bridge move to complete the circuit.
- The voltmeter measures the potential difference – a measure of the difference in ability of the two metals to lose electrons.
- Any two half-reactions can be paired and the cell potential measured.

Standard conditions

To make a fair comparison between two half-reactions, standard conditions are used:

1 Solutions are $1 \, mol \, dm^{-3}$.

2 Gases are at 100 kPa pressure.

3 The temperature is 298 K.

> When $E^\ominus$ values are used, half-reactions are always written as reductions. $E^\ominus$ values will be given on the examination paper.

Standard electrode (redox) potential, $E^\ominus$

To compare half-reactions, all half-reactions can be paired with the reaction

$2H^+ + 2e^- \rightarrow H_2$

The potential measured in this cell is $E^\ominus$.

The electrode for the hydrogen half-reaction is the standard hydrogen electrode (SHE).

In a pair of half-reactions, the one with more positive $E^\ominus$ value will occur, the one with the less positive $E^\ominus$ value will be reversed.

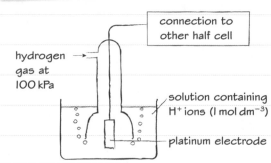

connection to other half cell

hydrogen gas at 100 kPa

solution containing H^+ ions ($1 \, mol \, dm^{-3}$)

platinum electrode

$E^\ominus$ for this half-reaction is defined as 0.00 V.

Worked example

How is the standard electrode potential of $Cl_2 + 2e^- \rightarrow 2Cl^-$ measured?   **(3 marks)**

An electrode is made by bubbling chlorine gas at 100 kPa into a solution containing $1 \, mol \, dm^{-3}$ chloride ions at 298 K. This electrode is connected to a SHE and the potential measured.

Now try this

For the redox reaction $Cl_2 + 2Br^- \rightarrow Br_2 + 2Cl^-$
(a) Write down the two half-reactions as reductions.
                                                        (2 marks)
(b) Which half-reaction has the more positive $E^\ominus$ value?
                                                        (1 mark)

> For the **standard** electrode potential, conditions must be specified.

Measuring and using standard electrode potentials

The **standard cell potential** can be calculated using the standard electrode potentials of the half-reactions.

🧪 Practical skills   **Measuring standard electrode potentials**

To measure the standard electrode potential the half-reaction is connected to a SHE.

1 For a metal/metal ions half-reaction, the metal (unless reactive with water) is used to make the electrode, which is dipped into a solution containing the ions at 1 mol dm^{-3}. Examples:

$Zn^{2+} + 2e^- \rightarrow Zn$ $Cu^{2+} + 2e^- \rightarrow Cu$

2 For a non-metal gas/gas ions half-reaction, the gas is bubbled at 100 kPa into a solution containing the ions at 1 mol dm^{-3}. Example (see previous page):

$2H^+ + 2e^- \rightarrow H_2$

3 For ions of the same element, the ions are mixed so that they are each 1 mol dm^{-3} and a platinum electrode is used. Example:

$Fe^{3+} + e^- \rightarrow Fe^{2+}$

Circuit to measure the standard electrode potential for $Fe^{3+} + e^- \rightarrow Fe^{2+}$

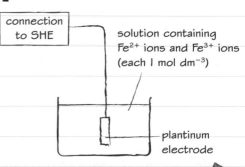

connection to SHE   solution containing Fe^{2+} ions and Fe^{3+} ions (each 1 mol dm^{-3})

platinum electrode

For electrodes containing a mixture of ions of the same element, they must be **equimolar** (of the same concentration); 1 mol dm^{-3} is just one example.

Standard electrode potential values

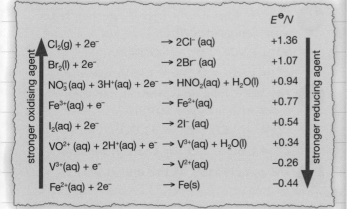

			$E^{\ominus}$/V
$Cl_2(g) + 2e^-$	$\rightarrow 2Cl^-(aq)$		+1.36
$Br_2(l) + 2e^-$	$\rightarrow 2Br^-(aq)$		+1.07
$NO_3^-(aq) + 3H^+(aq) + 2e^-$	$\rightarrow HNO_2(aq) + H_2O(l)$		+0.94
$Fe^{3+}(aq) + e^-$	$\rightarrow Fe^{2+}(aq)$		+0.77
$I_2(aq) + 2e^-$	$\rightarrow 2I^-(aq)$		+0.54
$VO^{2+}(aq) + 2H^+(aq) + e^-$	$\rightarrow V^{3+}(aq) + H_2O(l)$		+0.34
$V^{3+}(aq) + e^-$	$\rightarrow V^{2+}(aq)$		-0.26
$Fe^{2+}(aq) + 2e^-$	$\rightarrow Fe(s)$		-0.44

stronger oxidising agent ↑   stronger reducing agent ↓

Required values will be provided in the question.

Now try this

Fe^{3+} ions oxidise iodide ions to iodine.
Use the data above to:
(a) write the balanced ionic equation   **(1 mark)**
(b) calculate the $E^{\ominus}$ value for the reaction.   **(1 mark)**

Worked example

Using standard electrode potentials to predict the standard cell potential

Liquid bromine oxidises V^{3+} ions to VO^{2+} ions.

(a) Write the balanced ionic equation for this reaction.   **(1 mark)**

1 The half-equations are:
$Br_2 + 2e^- \rightarrow 2Br^-$
$VO^{2+} + 2H^+ + e^- \rightarrow V^{3+} + H_2O$

2 The second equation is doubled
$2V^{3+} + 2H_2O \rightarrow 2VO^{2+} + 4H^+ + 2e^-$

3 The half-equations are added and the electrons are cancelled
$Br_2 + 2V^{3+} + 2H_2O$
$\rightarrow 2Br^- + 2VO^{2+} + 4H^+$

(b) Use the standard electrode potentials to calculate the standard cell potential.   **(1 mark)**

$E^{\ominus} = +1.07 - 0.34 = +0.73\,V$

The second equation is reversed, so the sign of $E^{\ominus}$ is reversed, from +0.34V to −0.34V. $E^{\ominus}$ values are **not** multiplied even when the half-equation is (see page 104).

Predicting feasibility

The feasibility of a redox reaction can be predicted by using the standard electrode potentials of the two half-reactions, and hence the standard cell potential. This calculation applies whether or not the reaction is set up as a cell.

Predicting feasibility

- The method shown on page 107 can be used to calculate the standard cell potential.
- For a reaction to be feasible, $E^\ominus$ must be positive.
- A reaction with a negative standard cell potential will not happen, but the reverse reaction is feasible (and the $E^\ominus$ value is reversed).

Limitations in feasibility predictions

1. The $E^\ominus$ value will predict whether a reaction is feasible, but not the rate of reaction. Some reactions may be feasible but too slow to observe.

2. The $E^\ominus$ value is the standard cell potential, applying only under standard conditions (see page 106). Le Chatelier's principle (see page 52) can be used to determine the effect of non-standard conditions on $E^\ominus$.

Standard electrode potential values

		$E^\ominus$/V
$Cl_2(g) + 2e^-$	$\rightarrow 2Cl^-(aq)$	+1.36
$Br_2(l) + 2e^-$	$\rightarrow 2Br^-(aq)$	+1.07
$NO_3^-(aq) + 3H^+(aq) + 2e^-$	$\rightarrow HNO_2(aq) + H_2O(l)$	+0.94
$Fe^{3+}(aq) + e^-$	$\rightarrow Fe^{2+}(aq)$	+0.77
$I_2(aq) + 2e^-$	$\rightarrow 2I^-(aq)$	+0.54
$VO^{2+}(aq) + 2H^+(aq) + e^-$	$\rightarrow V^{3+}(aq) + H_2O(l)$	+0.34
$V^{3+}(aq) + e^-$	$\rightarrow V^{2+}(aq)$	-0.26
$Fe^{2+}(aq) + 2e^-$	$\rightarrow Fe(s)$	-0.44

Worked example

An excess of acidified potassium manganate(VII) was added to a solution containing $V^{2+}(aq)$ ions. Use the data above to calculate $E^\ominus$ values and determine the vanadium species present in the solution at the end of this reaction.   **(4 marks)**

1. The manganate(VII) half-reaction has an $E^\ominus$ value of +1.52 V.
   $V^{2+} \rightarrow V^{3+}$ $E^\ominus$ value +0.26V.
   $E^\ominus = 1.52 + 0.26$
   $= 1.78$ V: feasible.

2. The V^{3+} ions now present can then react with manganate(VII):
   $V^{3+} \rightarrow VO^{2+}$, $E^\ominus = -0.34$V
   $E^\ominus = 1.52 - 0.34 = 1.18$V: feasible.

3. The VO^{2+} ions now present can then react with manganate(VII):
   $VO^{2+} \rightarrow VO_2^+$, $E^\ominus = -1.00$V
   $E^\ominus = 1.52 - 1.00 = 0.52$V: feasible

4. The species left at the end is VO_2^+.

Worked example

1 mol dm^{-3} hydrogen peroxide solution reacts with 1 mol dm^{-3} iron(III) chloride solution at 25°C.

(a) Combine half-equations from above to write an equilibrium equation.   **(1 mark)**

$H_2O_2 + 2Fe^{3+} \rightarrow O_2 + 2H^+ + 2Fe^{2+}$

(b) Calculate the standard cell potential and state whether the reaction is feasible.   **(2 marks)**

$E^\ominus = -0.68 + 0.77 = +0.09$V

The reaction is feasible because $E^\ominus$ is positive.

(c) Explain how the E value could be increased.   **(2 marks)**

Increase the concentration of hydrogen peroxide or iron(III) ions, which would move the equilibrium to the right and increase E.

The question makes it clear that the hydrogen peroxide half-equation must be reversed. The iron half-equation is doubled so that each half-equation has two electrons.

In multi-step reactions, keep trying the next possible reaction until a negative $E^\ominus$ value is found.

Now try this

An excess of a solution containing iron(III) ions was added to a solution containing $V^{2+}(aq)$ ions. Use the data above to determine the vanadium species present in the solution at the end of this reaction.   **(2 marks)**

Storage and fuel cells

Redox reactions occur in storage cells and fuel cells in which the transfer of electrons happens through an external circuit. This provides a useful electric current.

Types of cell

Storage cells contain the substances required for the half reactions within the cell's casing. They are portable sources of electrical energy. A fuel cell is where the substances required for the reaction are provided continuously.

> More than one cell joined together (to give more power) is called *a battery* (although 'batteries' found in the shops may contain only one cell). Details of cells and the relevant electrode potentials will be provided in the exam.

Storage cell

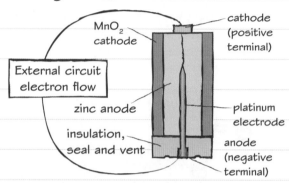

The **anode** is where oxidation occurs.
The **cathode** is where reduction occurs.

Half-reactions

$ZnO + H_2O + 2e^- \rightarrow Zn + 2OH^-$ $E^{\ominus} = -1.28V$

$2MnO_2 + H_2O + 2e^- \rightarrow Mn_2O_3 + 2OH^-$
$E^{\ominus} = +0.15 V$

Cell reaction

$Zn + 2MnO_2 \rightarrow ZnO + Mn_2O_3$
$E^{\ominus} = +1.28 + 0.15 = 1.43 V$

> In fuel cells the overall reaction is of the fuel (in this case, hydrogen) with oxygen from the air.

Comparison of cell types

• Storage cells are portable.
• If a lithium storage cell is damaged it can overheat and catch fire, and lithium compounds are toxic.
• The voltage from a storage cell drops over time (as the reactants are used up) but a fuel cell has constant voltage.
• Storage cells are difficult to dispose of but fuel cells produce only water.

> Make sure that the electrons balance before combining half-equations.

Fuel cell

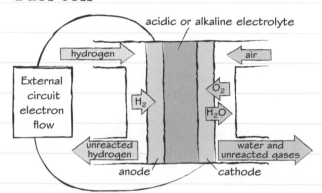

Half-reactions

① If the electrolyte is acidic:

$2H^+ + 2e^- \rightarrow H_2$ $E^{\ominus} = 0.00V$

$4H^+ + O_2 + 4e^- \rightarrow 2H_2O$ $E^{\ominus} = +1.23V$

Cell reaction

$2H_2 + O_2 \rightarrow 2H_2O; E^{\ominus} = 0 + 1.23 = 1.23V$

② If the electrolyte is alkaline:

$4H_2O + 4e^- \rightarrow 2H_2 + 4OH^-$
$E^{\ominus} = -0.83V$

$2H_2O + O_2 + 4e^- \rightarrow 4OH^-$
$E^{\ominus} = +0.40V$

> The $E^{\ominus}$ values for these two cells are the same because the overall reaction is the same.

Cell reaction

$2H_2 + O_2 \rightarrow 2H_2O; E^{\ominus} = 0 + 1.23 = 1.23V$

Worked example

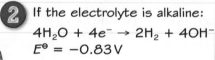

The half-equations are given for an ethanol fuel cell.

$12H^+ + 2CO_2 + 12e^- \rightarrow C_2H_5OH + 3H_2O$ $E^{\ominus} = +0.10V$

$4H^+ + O_2 + 4e^- \rightarrow 2H_2O$ $E^{\ominus} = +1.23V$

(a) Write the equation for the overall reaction.
                                                              (1 mark)

$C_2H_5OH + 3O_2 \rightarrow 2CO_2 + 3H_2O$

(b) Calculate $E^{\ominus}$.   **(2 marks)**

$E^{\ominus} = -0.10 + 1.23 = 1.13 V$

Now try this

Discuss potential environmental advantages of hydrogen fuel-cell powered vehicles.   **(3 marks)**

> Consider the source of the hydrogen fuel and the products of combustion.

Exam skills 9

This exam-style question uses knowledge and skills you have already revised. Look at pages 104, 108 and 100 for a reminder about redox, calculating the feasibility of reactions using electrode potentials and factors affecting lattice enthalpy.

Worked example

(a) Consider the electrode potentials given below.

$E^{\ominus}$/V

$$Mn^{2+}(aq) + 2e^- \rightleftharpoons Mn(s) \quad -1.18$$
$$Fe^{2+}(aq) + 2e^- \rightleftharpoons Fe(s) \quad -0.44$$
$$Ni^{2+}(aq) + 2e^- \rightleftharpoons Ni(s) \quad -0.25$$
$$Sn^{2+}(aq) + 2e^- \rightleftharpoons Sn(s) \quad -0.14$$
$$2H^+(aq) + 2e^- \rightleftharpoons H_2(g)$$

(i) State the value of the standard electrode potential of
$2H^+(aq) + 2e^- \rightleftharpoons H_2(g)$. **(1 mark)**

0 V

> This is the standard hydrogen electrode (SHE) and the standard electrode potential is exactly 0 V by definition – see page 106.

(ii) Write the ionic equation for the reaction when nickel reduces Fe^{2+}, and calculate the standard electrode potential. **(3 marks)**

$Ni(s) + Fe^{2+}(aq) \rightleftharpoons Ni^{2+}(aq) + Fe(s)$
$E^{\ominus} = +0.25 - 0.44 = -0.19$ V

> **Maths skills** The Ni $\rightleftharpoons$ Ni^{2+} + 2e$^-$ half-cell is the reverse of that given in the table (data is always given for reductions) so the $E^{\ominus}$ value is reversed.

(iii) If the two half cells for nickel and iron used in (ii) were connected under standard conditions, state and explain the reaction that would occur (if any) and the initial voltage of the cell. **(2 marks)**

The potential calculated in (ii) is negative, so the reverse reaction, $Ni^{2+}(aq) + Fe(s) \rightarrow Ni(s) + Fe^{2+}(aq)$, would occur giving a cell voltage of +0.19 V.

> Cell potential must be positive for a reaction to be feasible. As in this case, if the potential is negative, the reverse reaction is feasible.

(iv) Identify **all** of the species from the table that would reduce Sn^{2+}. **(1 mark)**

Ni, Fe and Mn.

> The Sn^{2+} reduction has a value of -0.14 V. An oxidation with a value of > +0.14 V is needed to give an overall $E^{\ominus}$ value that is positive.

(b) The lattice enthalpy of iron(II) chloride is -2631 kJ mol^{-1} and the lattice enthalpy of iron(III) chloride is -3865 kJ mol^{-1}.

(i) Write the equation for the reaction whose enthalpy change is the lattice enthalpy of iron(III) chloride. **(2 marks)**

$Fe^{3+}(g) + 3Cl^-(g) \rightarrow FeCl_3(s)$

> State symbols must be included in an equation for a reaction whose enthalpy change is the lattice enthalpy. This is because the definition of lattice enthalpy requires gaseous ions.

(ii) Explain, in terms of iron ions, why the lattice enthalpy for iron(III) chloride is more exothermic than the lattice enthalpy for iron(II) chloride. **(2 marks)**

The charge on the iron(III) ion is Fe^{3+}, higher than that on iron(II) ions. So there is a stronger electrostatic attraction between the Fe^{3+} ions and the chloride ions, and more energy is released on forming the solid.

(iii) Write an equation showing the reaction of iron(II) chloride with chlorine to form iron(III) chloride, and identify the oxidising agent in the reaction. **(2 marks)**

$2FeCl_2 + Cl_2 \rightarrow 2FeCl_3$
The oxidising agent is chlorine.

> The Fe^{2+} ions lose electrons to form Fe^{3+}, and are oxidised, so chlorine must be the oxidising agent. (Chlorine gains electrons and is reduced.)

The transition elements

The **transition elements** of Period 4 are those d-block elements that have an ion with an incomplete d sub-shell.

The d-block elements are those whose highest energy electron is in a d-orbital.

transition elements

						Hg												He
Li	Be											B	C	N	O	F	Ne
Na	Mg											Al	Si	P	S	Cl	Ar
K	Ca	Sc	Ti	V	Cr	Mn	Fe	Co	Ni	Cu	Zn	Ga	Ge	As	Se	Br	Kr
Rb	Sr	Y	Zr	Nb	Mo	Tc	Ru	Rh	Pd	Ag	Cd	In	Sn	Sb	Te	I	Xe
Cs	Ba	La	Hf	Ta	W	Re	Os	Ir	Pt	Au	Hg	Tl	Pb	Bi	Po	At	Rn
Fr	Ra	Ac															

'd-block' elements

Electronic configurations

Scandium's ion (Sc^{3+}) has an empty 3d sub-shell and zinc's ion (Zn^{2+}) has a full 3d sub-shell, so these elements are **not** transition elements.

> When d-block elements in this period form ions, they always lose their 4s electrons first.

scandium forms 3+ ions, with electronic configuration [Ar]

the transition elements in this period form ions by losing one or more electrons, leaving an incomplete 3d sub-shell

chromium atoms and copper atoms have only one electron in the 4s orbital

zinc 2+ ions, with electronic configuration [Ar], $3d^{10}$

		3d						4s
Sc	[Ar]	↑						↑↓
Ti	[Ar]	↑	↑					↑↓
V	[Ar]	↑	↑	↑				↑↓
Cr	[Ar]	↑	↑	↑	↑	↑		↑
Mn	[Ar]	↑	↑	↑	↑	↑		↑↓
Fe	[Ar]	↑↓	↑	↑	↑	↑		↑↓
Cu	[Ar]	↑↓	↑↓	↑	↑	↑		↑↓
Ni	[Ar]	↑↓	↑↓	↑↓	↑	↑		↑↓
Cu	[Ar]	↑↓	↑↓	↑↓	↑↓	↑↓		↑
Zn	[Ar]	↑↓	↑↓	↑↓	↑↓	↑↓		↑↓

Variable oxidation states of transition elements

- Transition elements form ions with different charges, so they have different oxidation states.
- Iron can be, for example, Fe(II) [Fe^{2+}] or Fe(III) [Fe^{3+}].
- Copper can be Cu(I) [Cu^+] or Cu(II) [Cu^{2+}].
- Chromium can be, for example, Cr(III) [Cr^{3+}] or Cr(VI) [$Cr_2O_7^{2-}$].
- Redox reactions can be used to convert between these oxidation states.

> Metals in Groups 1, 2 and 3 always form ions with a charge 1+, 2+ and 3+ respectively. Born–Haber cycles (see page 99) show that ionic compounds with these ions are the most energetically favourable.

Worked example

(a) Write the full electronic configurations of the following atoms and ions:

     Sc, Sc^{3+}, Ti^{2+}, V^{3+}, Fe^{3+} **(5 marks)**

Sc	$1s^2 2s^2 2p^6 3s^2 3p^6 3d^1 4s^2$
Sc^{3+}	$1s^2 2s^2 2p^6 3s^2 3p^6$
Ti^{2+}	$1s^2 2s^2 2p^6 3s^2 3p^6 3d^2$
V^{3+}	$1s^2 2s^2 2p^6 3s^2 3p^6 3d^2$
Fe^{3+}	$1s^2 2s^2 2p^6 3s^2 3p^6 3d^5$

> Look up the atomic numbers on the Periodic Table. The 4s orbital is written after the 3d orbital. Remember that the 4s electrons are lost first.

(b) Explain whether scandium is a transition element.   **(1 mark)**

Scandium is not a transition element, because its only ion, Sc^{3+}, has an empty 3d sub-shell.

Now try this

Write the electronic configurations for Cr, Cr^{3+}, Zn and Zn^{2+}. Use these to explain why chromium and zinc are d-block elements but only chromium is a transition element.   **(4 marks)**

Properties of transition elements

The transition elements have characteristic properties that distinguish them from other metallic elements.

Transition elements have coloured ions

Colour	Aqueous ions	Solids (precipitates)
green	$[Fe(H_2O)_6]^{2+}$ $[Cr(H_2O)_6]^{3+}$ $[Cr(OH)_6]^{3-}$	$Fe(OH)_2$* $Cr(OH)_3$
violet		$[Cr(NH_3)_6]^{3+}$
brown		$Fe(OH)_3$
blue	$[Cu(H_2O)_6]^{2+}$ (pale) $[Cu(NH_3)_4(H_2O)_2]^{2+}$ (dark)	$Cu(OH)_2$
pink	$[Mn(H_2O)_6]^{2+}$	
white		*$Mn(OH)_2$ Cu(I) halides
yellow	$[Fe(H_2O)_6]^{3+}$ $[CuCl_4]^{2-}$	
orange	$Cr_2O_7^{2-}$	

*precipitate darkens in air

In contrast, compounds of non-transition metals have white compounds that, if soluble, form colourless solutions.

Transition elements as catalysts

- Transition elements have variable oxidation states (see page 111).
- This allows transition elements and their compounds to act as catalysts.

Transition metal catalysts in industry

- Many industrial processes use catalysts that are transition metal elements or their compounds.
- This reduces the temperature at which a reaction is carried out.
- Energy costs and the impact of energy use on the environment are reduced.
- However, many transition metals are toxic and must be used with care.

Practical skills — Demonstrating catalytic activity

Hydrogen peroxide solution decomposes.

$$2H_2O_2(aq) \rightarrow 2H_2O(l) + O_2(g)$$

Manganese(IV) oxide, MnO_2, can be shown to be a catalyst for this reaction.

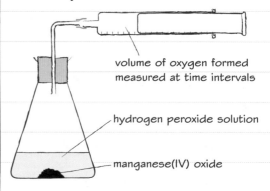

volume of oxygen formed measured at time intervals

hydrogen peroxide solution

manganese(IV) oxide

1 The rate of reaction can be measured (see page 48) with and without manganese(IV) oxide.

2 After the reaction the manganese(IV) oxide can be filtered off, washed and dried to show it has not lost mass.

Worked example

Solutions of Zn^{2+} ions and Cu^{2+} ions both form precipitates when sodium hydroxide solution is added.

(a) Write the ionic equation for the formation of copper(II) hydroxide. **(1 mark)**

$$Cu^{2+}(aq) + 2OH^-(aq) \rightarrow Cu(OH)_2(s)$$

(b) State the colours of zinc hydroxide and copper(II) hydroxide. **(1 mark)**

Zinc hydroxide is white and copper(II) hydroxide is blue.

(c) State how the position of zinc and of copper in the Periodic Table is linked to the colours of the hydroxides. **(1 mark)**

Zinc is not a transition metal so has white compounds; copper is a transition metal and has coloured compounds.

Now try this

Zinc reacts with hydrochloric acid:
$$Zn + 2HCl \rightarrow ZnCl_2 + H_2$$
Cu^{2+} ions act as a catalyst for this reaction.
Design an experiment to prove that Cu^{2+} ions greatly increase the rate of reaction. Give an outline of the method and the measurements you would take. **(4 marks)**

Complex ions

A **complex ion** is formed when **ligands** join to a metal ion or atom by **coordinate** bonds.

$[Cr(NH_3)_6]^{3+}$ complex ion

- The nitrogen on an ammonia molecule has a lone pair of electrons, which is donated to the Cr^{3+} forming a coordinate bond (for coordinate bonding see page 20).
- Ammonia is the ligand.
- The ligands form six coordinate bonds – the **coordination number** is 6.

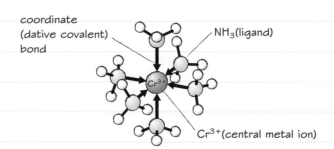

coordinate (dative covalent) bond

NH_3(ligand)

Cr^{3+}(central metal ion)

Monodentate and bidentate ligands

- Monodentate ligands form one coordinate bond.
- Examples of monodentate ligands include H_2O, Cl^- and NH_3.
- Bidentate ligands form two coordinate bonds – so they must have two atoms with a lone pair.
- An example of a bidentate ligand is 1,2-diaminoethane ('en'), $NH_2CH_2CH_2NH_2$.
- en forms coordinate bonds using the lone pair on each of its N atoms.

Coordination number

- The coordination number is the number of coordinate bonds formed by the ligands.
- The coordination number can be 2, 4 or 6.
- Examples of six-fold coordination are on this page, and four-fold coordination on the next page.

The coordination number is not the number of **ligands**, but the number of **coordinate bonds**.

Examples of octahedral complex ions

$[Fe(H_2O)_6]^{2+}$

hexaaquairon(II) ion

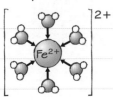

monodentate ligand
coordination number 6

$[Fe(H_2O)_6]^{3+}$

hexaaquairon(III) ion

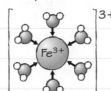

monodentate ligand
coordination number 6

$[Co(NH_2CH_2CH_2NH_2)_3]^{3+}$

bidentate ligand
coordination number 6

An octahedral shape can only form if there are six coordinate bonds.

Worked example

The complex ion $[Ag(NH_3)_2]^{2+}$ has:

A   coordination number 2, monodentate ligands
B   coordination number 2, bidentate ligands
C   coordination number 6, monodentate ligands
D   coordination number 6, bidentate ligands. **(1 mark)**

A

Now try this

Draw the complex ion $[Fe(H_2O)_6]^{2+}$ showing the coordinate bonds.
State and explain the shape of this ion.   **(4 marks)**

Each ammonia forms one coordinate bond (each N atom has one lone pair) making the ligand monodentate. There are two coordinate bonds in total – making the coordination number 2.

4-fold coordination and isomerism

Complexes with four-fold coordination may be planar or tetrahedral. Some four-fold and six-fold complexes exhibit stereoisomerism.

Tetrahedral complexes

Many four-coordinate complexes have a tetrahedral shape, for example $CuCl_4^{2-}$ and $CoCl_4^{2-}$.

Square planar complexes

Some four-coordinate complexes have a square planar shape, for example $Pt(NH_3)_2Cl_2$ (platin).

Stereoisomerism:

cis–trans isomerism

- Some four-fold and six-fold coordinate complexes can exist as *cis-trans* isomers.
- Platin is an example, shown below.
- Cis-platin is an anti-cancer drug which binds to the DNA of cancerous cells and prevents cell division; however, it does have unpleasant side effects.

Examples of four-coordinate complex ions

$CuCl_4^{2-}$

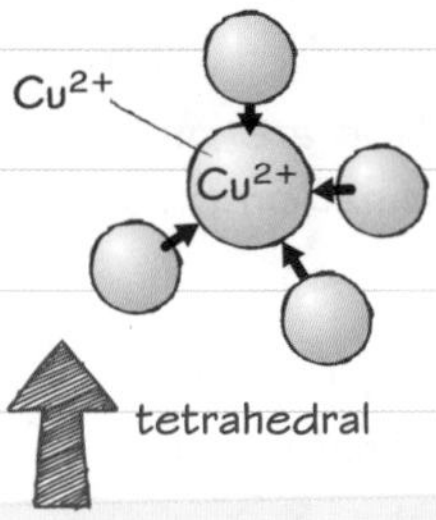

Cu²⁺

tetrahedral

cis-platin, $Pt(NH_3)_2Cl_2$

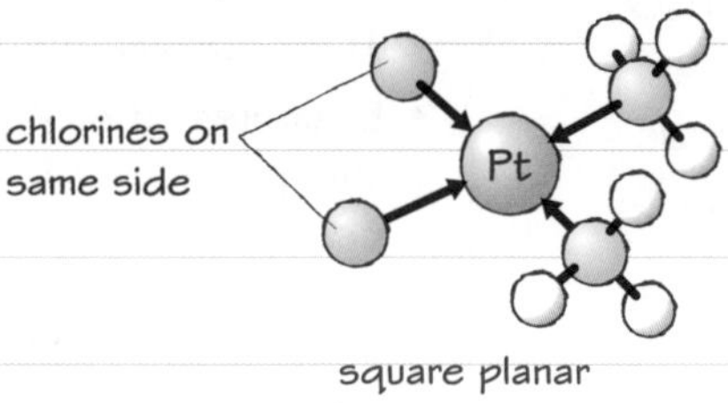

chlorines on same side

square planar

trans-platin, $Pt(NH_3)_2Cl_2$

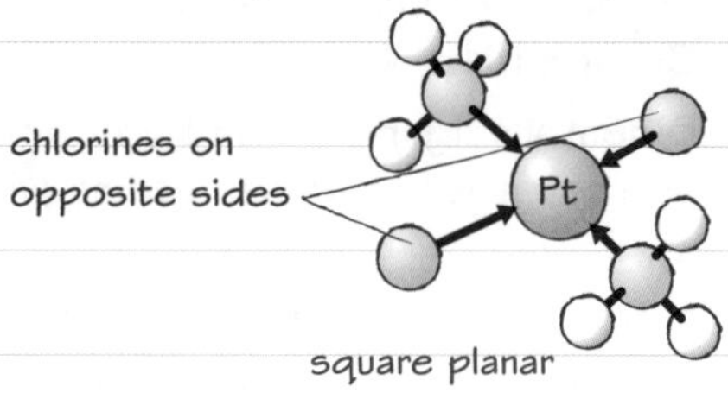

chlorines on opposite sides

square planar

$CoCl_4^{2-}$ has same shape as $CuCl_4^{2-}$.

Complexes whose central metal atom/ion has an electronic configuration d^8 (for example, Pt and Ni^{2+}) are typically square planar.

Stereoisomerism: optical isomerism in complexes

- Some six-fold coordinate complexes can exist as optical isomers.
- These are possible with bidentate ligands.
- An example is $[Ni(NH_2CH_2CH_2NH_2)_3]^{2+}$.

Remember that optical isomers are non-superimposable mirror images (see page 133).

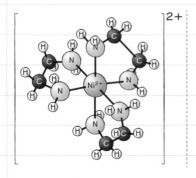

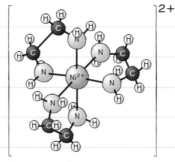

Be careful! Choose the ligand that appears twice – in this case Cl.
Draw the *cis* isomer so that the Cl ligands appear next to each other.
The *trans* isomer has the Cl ligands opposite each other.

Worked example

Draw diagrams showing *cis-trans* isomers of $[Cr(NH_3)_4Cl_2]^+$.   **(2 marks)**

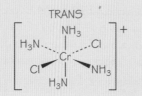

CIS   TRANS

Now try this

Which of the following complexes exhibit *cis-trans* isomerism?   **(1 mark)**

A $CuCl_4^{2-}$

B $[Cu(H_2O)_6]^{2+}$

C $Ni(NH_3)_2Cl_2$

D $[Ag(NH_3)_2]^+$

Precipitation reactions

Solutions of transition element ions form coloured precipitates with sodium hydroxide solution or ammonia solution, and these precipitates can be used as tests for these ions.

Aqua ions

The transition metal ions Cu^{2+}, Fe^{2+}, Fe^{3+}, Mn^{2+}, Cr^{3+} all form hexaaqua ions in solution, $[M(H_2O)_6]^{n+}$.

The aqueous M^{n+} ion may also be written as $M^{n+}(aq)$.

The precipitate may also be written as $M(OH)_n$

In these **acid–base** reactions, the hydroxide ions remove an H^+ from a water ligand (leaving the ligand as an OH^- ion), and forming H_2O (in contrast to the reactions on the next page).

Sodium hydroxide and ammonia solutions

- $NaOH(s) + aq \rightarrow Na^+(aq) + OH^-(aq)$
- $NH_3(aq) + H_2O(l) \rightleftharpoons NH_4^+(aq) + OH^-(aq)$
- Transition metal ions in solution react with the hydroxide ions in either solution to form a neutral precipitate:

$M^{n+}(aq) + nOH^-(aq) \rightarrow M(OH)_n(s)$

The three n values are equal.

Precipitates formed by dropwise addition of hydroxide ions

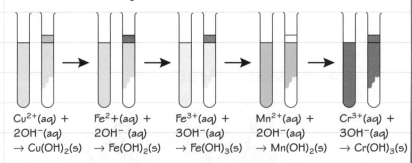

$Cu^{2+}(aq) +$ $Fe^{2+}(aq) +$ $Fe^{3+}(aq) +$ $Mn^{2+}(aq) +$ $Cr^{3+}(aq) +$
$2OH^-(aq)$ $2OH^-(aq)$ $3OH^-(aq)$ $2OH^-(aq)$ $3OH^-(aq)$
$\rightarrow Cu(OH)_2(s)$ $\rightarrow Fe(OH)_2(s)$ $\rightarrow Fe(OH)_3(s)$ $\rightarrow Mn(OH)_2(s)$ $\rightarrow Cr(OH)_3(s)$

The effect of excess sodium hydroxide solution or ammonia solution

1 Fe^{2+}, Fe^{3+}, Mn^{2+}
Excess sodium hydroxide or ammonia solution has no effect on the precipitate formed.

2 Cr^{3+}
Excess sodium hydroxide or ammonia solutions cause the initial $Cr(OH)_3$ precipitate to dissolve to form a green $[Cr(OH)_6]^{3-}$ solution or a violet $[Cr(NH_3)_6]^{3+}$ solution.
$Cr(OH)_3 + 3OH^- \rightarrow [Cr(OH)_6]^{3-}$
$Cr(OH)_3 + 6NH_3 \rightarrow [Cr(NH_3)_6]^{3+} + 3OH^-$

3 Cu^{2+}
Excess sodium hydroxide solution has no effect on the initial $Cu(OH)_2$ precipitate.
Excess ammonia solution causes the initial $Cu(OH)_2$ precipitate to dissolve to make a deep blue solution.
$Cu(OH)_2(H_2O)_4 + 4NH_3 \rightarrow$
$\quad [Cu(NH_3)_4(H_2O)_2]^{2+} + 2OH^{3-} + 2H_2O$

Tests for ions

🧪 **Practical skills**

1 Cu^{2+}, Fe^{2+}, Fe^{3+}, Mn^{2+}, Cr^{3+}
- make a solution of the salt
- add sodium hydroxide solution or ammonia solution dropwise, with shaking
- observe the precipitate formed
- add excess sodium hydroxide solution or ammonia solution to see whether the precipitate dissolves
- the expected results are given on this page.

2 CO_3^{2-}, Cl^-, Br^-, I^-, SO_4^{2-}, NH_4^+
- the tests for these ions are given on page 37.

Worked example

State what is observed when sodium hydroxide solution is added to a solution of Fe^{2+} ions and explain why the observation changes on standing.

(3 marks)

In the green solution a green precipitate forms, which darkens because the $Fe(OH)_2$ is oxidised to $Fe(OH)_3$ by oxygen in the air.

Now try this

How would you distinguish between two green solutions: Fe^{2+} and Cr^{3+}? **(3 marks)**

115

Ligand substitution reactions

Transition element complex ions can change their ligands, and these reactions are often accompanied by a colour change.

Ligand substitution reactions of copper and chromium

1 When ammonia solution is added to $Cu^{2+}(aq)$ or $Cr^{3+}(aq)$ some water ligands are substituted by ammonia ligands.

2 When concentrated hydrochloric acid is added to $Cu^{2+}(aq)$ some water ligands are substituted by chloride ligands.

In these **substitution** reactions, ligands are replaced, in contrast to the acid–base reactions on the previous page.

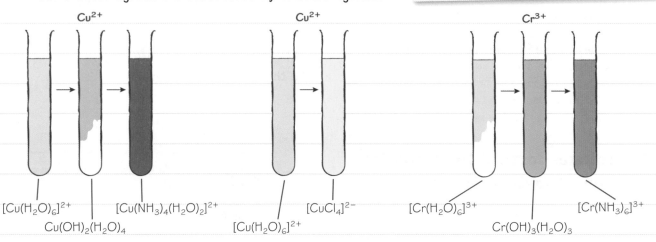

Cu^{2+}

Cu^{2+}

Cr^{3+}

$[Cu(H_2O)_6]^{2+}$ $[Cu(NH_3)_4(H_2O)_2]^{2+}$

$Cu(OH)_2(H_2O)_4$

$[CuCl_4]^{2-}$

$[Cu(H_2O)_6]^{2+}$

$[Cr(H_2O)_6]^{3+}$ $[Cr(NH_3)_6]^{3+}$

$Cr(OH)_3(H_2O)_3$

Worked example

A sulfate transition metal containing **A** is dissolved in water to form a blue solution containing aqua ions, **B**. When ammonia solution is added dropwise to the solution, a pale blue precipitate **C** forms, and with excess ammonia the precipitate dissolves to form a deep blue solution, **D**. If concentrated hydrochloric acid is added to **B** the solution turns green then yellow. The yellow solution contains the complex ion **E**.

Give the formulae of the species **A**, **B**, **C**, **D** and **E**, and write an ionic equation for **A → B**.   **(6 marks)**

A $CuSO_4$

B $[Cu(H_2O)_6]^{2+}$

C $[Cu(OH)_2(H_2O)_4]$

D $[Cu(NH_3)_4(H_2O)_2]^{2+}$

E $[CuCl_4]^{2-}$

$$CuSO_4(s) + 6H_2O(l) \rightarrow [Cu(H_2O)_6]^{2+} + SO_4^{2-}(aq)$$

Iron in haemoglobin

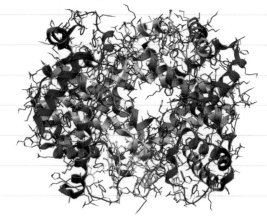

- Oxygen in the lungs (the ligand) binds to an Fe^{2+} ion in haemoglobin in a red blood cell.

- Blood is carried to parts of the body where the oxygen is released to take part in respiration.

- If CO is present, haemoglobin undergoes ligand substitution replacing oxygen with CO, preventing oxygen transport to the body and causing death.

Now try this

(a) Considering the information above, give ionic equations for the changes **B → C**, **C → D**, **B → E**.   **(3 marks)**

(b) Explain why in the change **B → E** the mixture turns green before turning yellow.   **(1 mark)**

Think about what a mixture of **B** and **E** would look like.

Redox reactions of transition elements

Transition elements can readily change their oxidation state, and these reactions are often accompanied by a colour change.

Redox reactions of iron, chromium and copper

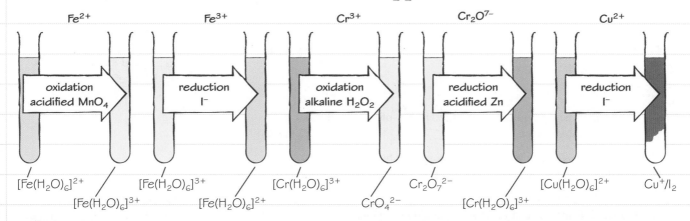

Fe^{2+} Fe^{3+} Cr^{3+} $Cr_2O_7^{-}$ Cu^{2+}

oxidation acidified MnO_4 reduction I^- oxidation alkaline H_2O_2 reduction acidified Zn   reduction I^-

$[Fe(H_2O)_6]^{2+}$ $[Fe(H_2O)_6]^{3+}$ $[Cr(H_2O)_6]^{3+}$ $Cr_2O_7^{2-}$ $[Cu(H_2O)_6]^{2+}$ Cu^+/I_2

$[Fe(H_2O)_6]^{3+}$ $[Fe(H_2O)_6]^{2+}$ CrO_4^{2-} $[Cr(H_2O)_6]^{3+}$

1 Fe^{2+} and Cr^{3+} can be oxidised by the reagents shown.

2 Fe^{3+}, $Cr_2O_7^{2-}$ and Cu^{2+} can be reduced by the reagents shown.

> You do not have to recall equations for these reactions, but you should be able to construct them using half-equations and oxidation numbers (see page 104).

Disproportionation of Cu^+

Another redox reaction of copper is the disproportionation (see page 35) of Cu^+.

- Copper(I) compounds are white solids.
- When soluble copper(I) compounds are dissolved in water, disproportionation occurs.

$$2Cu^+ \rightarrow Cu^{2+} + Cu$$

oxidation number: +1   +2   0

- Half of the Cu^+ ions are oxidised to Cu^{2+} and half are reduced to Cu.
- When a white copper(I) compound is shaken with water, a blue solution of $Cu^{2+}(aq)$ is formed with a red-brown precipitate of Cu.

— $Cu^{2+}(aq)$

— Cu

Worked example

(a) Write a half-equation for the reduction of $Cr_2O_7^{2-}$ to Cr^{3+} in acidic solution. **(2 marks)**

$Cr_2O_7^{2-} + 14H^+ + 6e^- \rightarrow 2Cr^{3+} + 7H_2O$

(b) What would be observed in the reaction above? **(2 marks)**

An orange solution forms a green solution.

> The $Cr_2O_7^{2-}$ is orange and Cr^{3+} is green. You meet this when alcohols and aldehydes are oxidised by $Cr_2O_7^{2-}$ (see page 68).

Now try this

Consider the following half equations:
$$MnO_4^- + 8H^+ + 5e^- \rightarrow Mn^{2+} + 4H_2O$$
$$Fe^{2+} \rightarrow Fe^{3+} + e^-$$

(a) Write the balanced ionic equation for the oxidation of Fe^{2+} by acidified potassium manganate(VII) solution. **(1 mark)**

(b) Describe all of the colour changes that occur. **(2 marks)**

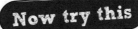

Consider both the iron ions and the manganate ions.

Exam skills 10

This exam-style question uses knowledge and skills you have already revised. Look at pages 111 and 112 for a reminder about transition elements and their properties.

Worked example

The atomic number of scandium is 21, and a scandium atom has an electronic configuration $[Ar]3d^14s^2$. The atomic number of cobalt is 27.

(a) (i) Give the electronic configuration of a cobalt atom and a Co^{2+} ion.

(2 marks)

Co $[Ar]\ 3d^74s^2$
$Co^{2+}\ [Ar]\ 3d^7$

> In electronic configurations, the previous Group 0 element in square brackets can be used to indicate the electrons up to the last full shell.

(ii) Explain why scandium is classified as a d-block element but not a transition element. **(2 marks)**

Scandium is a d-block element because its highest energy electron is in a 3d orbital. It is not a transition metal because its ion, Sc^{3+}, has no d electrons.

(b) Cobalt(II) chloride can react with the bidentate ligand $NH_2CH_2CH_2NH_2$ (en).

(i) What is the IUPAC name of en? **(1 mark)**

1,2-diaminoethane

(ii) What is the meaning of the term *bidentate*? **(1 mark)**

A ligand that forms two coordinate bonds with the central atom.

(c) The formula of a compound that can be formed is $[CoCl_2en_2]Cl$.

(i) What is the coordination number of cobalt in this complex ion? **(1 mark)**

6

> The complex ion is $[CoCl_2en_2]^+$, but in a compound the positive complex ion will be associated with an anion, called the **counter-ion**, in this case, the last Cl^- ion. This ion is not part of the complex ion (it is **not** bonded to the Co, but ionically attracted to the complex ion as a whole). $NH_2CH_2CH_2NH_2$ is neutral, and the chloride ions have a 1− charge, so Co must be Co^{3+} to give an overall charge on the complex ion of 1+.

(ii) By considering the oxidation state of cobalt in the complex ion, explain whether the reaction of cobalt(II) chloride and $NH_2CH_2CH_2NH_2$ to form this complex is a redox reaction.

(2 marks)

In the complex cobalt has an oxidation state of +3. Cobalt's oxidation state has increased from +2 to +3 which is oxidation, so this is a redox reaction.

(iii) A solution containing 0.100 mol of the compound $[CoCl_2en_2]$ Cl has an excess of silver nitrate solution added. The white precipitate formed is filtered off, washed and dried. The mass of the precipitate is 14.3 g. Identify the precipitate and then use the data to explain the mass of precipitate formed. **(3 marks)**

> **Maths skills** If you are not sure what to do in a question such as this, calculate the number of moles of each substance. Once you have done this, you can see that 1 mol of complex only gives 1 mol of Cl^- ions (and hence 1 mol of AgCl precipitate).

The white precipitate is silver chloride.

$M_r(AgCl) = 107.9 + 35.5 = 143.4$

mol of AgCl $= \dfrac{14.3}{143.4} = 0.0997$ mol

mol of complex = 0.100, so a 1:1 ratio.

The two chloride ligands are attached to the cobalt and do not form a precipitate, so only one Cl^- ion forms the precipitate.

The bonding in benzene rings

Benzene, C_6H_6, has a cyclic bonding arrangement first proposed by Kekulé and developed by others in light of more evidence.

Kekulé's model

Kekulé proposed a structure for benzene based on alternate double and single bonds. However, many chemists at the time challenged Kekulé's ideas, as his structure did not explain the low reactivity of benzene. Other chemists pointed out that if Kekulé was correct, benzene should react with bromine water in the same way as alkenes. Experiments showed that there is *no reaction*.

Kekulé's model evolves

The Kekulé model was amended to explain benzene's unreactivity by suggesting that a benzene molecule resonated rapidly between two forms of the structure. This would give bromine no chance to react with the double bond.

However, over time more evidence was uncovered that challenged the Kekulé model. For instance, benzene was found to have:

- six carbon-carbon bonds of equal length that are shorter than the C-C bond but longer than C=C
- a value for the enthalpy of hydrogenation significantly less exothermic than three times that of cyclohexene.

The delocalised model of benzene

By considering the way the electron orbitals overlap in benzene a structure was proposed that matches the evidence from bond lengths and the enthalpy of hydrogenation.

Each carbon atom forms a sigma bond to two other carbon atoms and a hydrogen atom. This forms the planar hexagonal ring. The remaining electron from each carbon is in a p orbital. The p orbitals overlap sideways to form a ring of electron density in pi bonds above and below the planar hexagonal ring.

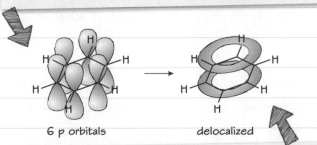

6 p orbitals   delocalized

The 6 electrons in the pi bonds spread over the whole ring. They are **delocalised** electrons. This is shown in the skeletal formula of benzene as a ring.

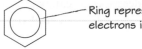
Ring represents delocalised electrons in pi bonds

The delocalisation means that this structure is more stable than Kekule's model.

Worked example

Explain the differences in the enthalpies of hydrogen for the two possible structures of benzene (as shown on the diagram).

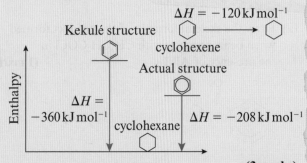

(3 marks)

The calculated value for the Kekulé structure of benzene is -360 kJ mol^{-1}, which is three times that of cyclohexene as the proposed has three carbon-carbon double bonds.

The actual value for benzene is smaller in magnitude (-208 kJ mol^{-1}), as its structure is more stable, due to delocalisation of the electrons in the pi bonds.

Now try this

Would you expect benzene to react with chlorine at room temperature and pressure? Explain your answer.

(2 marks)

Reactions of benzene rings

Compounds containing benzene rings are known as aromatic compounds. The delocalised electrons in the ring make it quite stable but benzene will react with some species in electrophilic substitution reactions, provided the attacking species is electrophilic enough.

The reactions of benzene

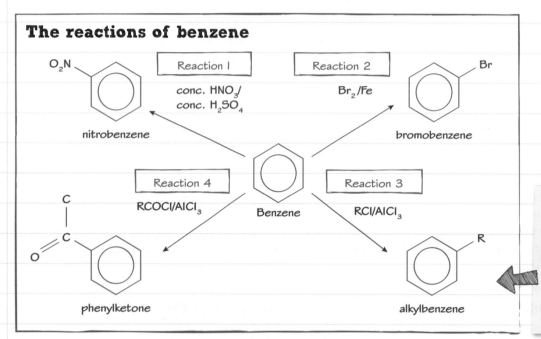

Note that reactions 3 and 4 are often referred to Friedal-Crafts Reactions after the two chemists who originally worked on them.

Worked example

(a) Draw the skeletal formula of the product formed when benzene reacts with CH_3CH_2COCl in the presence of $AlCl_3$. **(1 mark)**

(b) Draw the skeletal formula of the product formed when benzene reacts with $CH_3CH(Cl)CH_3$ in the presence of $AlCl_3$. **(1 mark)**

(c) Suggest the conditions needed for reaction 1. Justify your answer. **(2 marks)**

Keep temperature below 50°C. Higher temperatures increase the chance of substituting more than one nitro group onto the ring.

Now try this

The role of $AlCl_3$ in the electrophilic substitution reaction of chlorine with benzene is often described as catalytic. Explain, using equations, why only a small amount of $AlCl_3$ is needed in such reactions. **(2 marks)**

Electrophilic substitution reactions

Benzene and compounds with benzene rings, or arenes, can all be attacked by a variety of electrophiles. Although the electrophiles are generated in a number of ways, the mechanism is the same in each case.

Electrophilic substitution mechanism

The stages of the mechanism are:

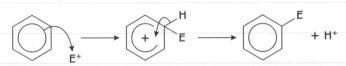

1 The electrophile accepts an electron pair from the delocalised electrons in the ring and forms a covalent bond.

2 A positive intermediate forms. It is unstable as the delocalised ring has been disrupted.

3 An electron pair from the C–H bond returns to the ring, making the ring stable once again. This releases an H^+ ion.

Electrophiles and products

The electrophiles that you need to recognise for your exam, and the products they will form in the reaction with a benzene ring, are in the table.

Electrophile	Product
NO_2^+	nitrobenzene
X^+ (halogen)	halobenzene
R^+ (alkyl group)	alkylbenzene
$(RCO)^+$ (acyl group)	phenylketone

The compound 4-nitrophenol can be formed from the reaction of 4-chlorophenol with nitric acid.

OH
(benzene ring) $\xrightarrow{HNO_3}$ OH (benzene ring)
Cl NO_2

Draw the mechanism for the reaction.   **(4 marks)**

OH (ring) → OH (ring with +) → OH (ring)   + Cl^+
Cl NO_2^+ Cl NO_2 NO_2

Don't let unusual compounds catch you off guard in this type of question. The mechanism is the same for all electrophilic substitution reactions. The only difference in this example is that a chlorine is substituted not a hydrogen – you're not expected to know this so you're told it in the question!

Be careful!

Make sure you know the precise details of mechanisms:

- The arrow to the electrophile starts on the ring.
- The gap in the ring of the intermediate is facing the carbon attacked by the electrophile.
- The gap does *not* stretch beyond the carbons next to carbon that is attacked.
- The arrow from the C–Cl bond points into the gap to re-form the ring.
- To include the formula of the species released from the ring, often H^+.

Now try this

Draw a suggested mechanism for the reaction of methylbenzene with sulfuric acid. You can assume the electrophile formed from sulfuric acid has the formula $^+SO_3$.   **(3 marks)**

Comparing the reactivity of alkenes and aromatic compounds

Both alkenes and aromatic compounds will react with electrophiles. As the pi bond in alkenes is more electron dense they will be more reactive with electrophiles.

How ethene and benzene interact with bromine

As a molecule of bromine approaches ethene or benzene, what happens is determined by the electron clouds due to the pi bonds present in the two hydrocarbons.

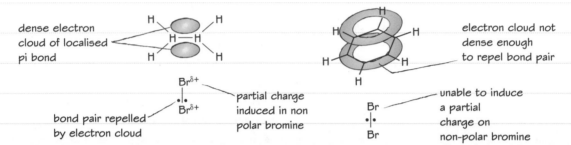

dense electron cloud of localised pi bond

electron cloud not dense enough to repel bond pair

$Br^{\delta+}$
$Br^{\delta+}$
partial charge induced in non polar bromine

bond pair repelled by electron cloud

unable to induce a partial charge on non-polar bromine

Br
Br

In ethene	In benzene
✓ High electron density in localised pi bond repels bond pair in Br_2.	✓ Electron density in pi bonds is delocalised over the ring, so less dense than in ethane.
✓ Partial charge is induced in Br_2.	✓ Bond pair in Br_2 *not* significantly repelled.
✓ Electrophilic bromine atom accepts electron pair to form bond to carbon.	✓ No partial charge induced in Br_2.
	✓ Non-polar bromine cannot react with benzene.

In order to get non-polar bromine to react with benzene, a catalyst is used, for example Fe. Often called a 'halogen carrier', Fe reacts with Br_2 to form an electrophile, Br^+. This is sufficiently electrophilic to interact with the delocalised electrons in benzene.

$2Fe + 3Br_2 \rightarrow 2FeBr_3$

$FeBr_3 + Br_2 \rightarrow FeBr_4^- + Br^+$ —— electrophile

Worked example

Methyl groups tend to release electron density towards neighbouring atoms. Use this idea, and your knowledge of electrophilic substitution, to compare the reactivity of methylbenzene and benzene with bromine.   **(4 marks)**

• The methyl group releases electrons into the ring.
• This increases the electron density of the ring, making it more attractive to electrophiles.
• So it is more likely to induce a positive charge on the bromine molecule.
• Hence methylbenzene is likely to be more reactive with bromine than benzene.

Although you are not expected to remember the effect of the methyl groups on electrophilic substitution reactions, this question gives you enough information to apply your specification knowledge to the problem.

Now try this

Describe and explain what you would *see* when samples of ethene and benzene are gently shaken with bromine water.   **(3 marks)**

Phenol

Phenols are molecules containing a benzene ring with a hydroxyl group attached. This group gives phenols some different properties to benzene, such as their acidity and their increased ease of substitution with electrophiles.

The acidity of phenol

Phenols can behave as an acid and donate a hydrogen ion from the hydroxyl group to strong bases.

Reaction with sodium hydroxide

However, phenol is *not* acidic enough to react with weak bases such as carbonates.

phenol $\xrightarrow{+NaOH}$ sodium phenoxide $+H_2O$

phenol   sodium phenoxide

Reactions of phenol

Phenol takes part in electrophilic substitutions much more readily than benzene.

2,4,6-tribromophenol $\xleftarrow{Br_2}$ phenol $\xrightarrow[HNO_3]{dilute}$ 2-nitrophenol and 4-nitrophenol

The reaction with bromine forms the trisubstituted product.

The –OH group takes priority when numbering substituted groups on the phenol ring. Hence the carbon attached to the –OH group is carbon number 1. The other groups attached to the ring are numbered accordingly.

Reactivity of phenol

Phenol is more reactive with electrophiles than benzene because of the higher electron density in its delocalised ring of pi electrons.

Non-bonding, or a lone, pair of electrons are drawn into delocalised ring of pi electrons

This increases electron density of the ring in phenol, compared to benzene

Notice that the conditions needed for phenol to react are less harsh than those needed for benzene. In phenol's reaction with Br_2 no halogen carrier is required and in the nitration only dilute nitric acid is required.

Describe how you could use bromine to distinguish between phenol and benzene. **(2 marks)**

Add bromine water to each and observe any changes. If the substance is phenol a white precipitate forms. With benzene there would be no change.

> This is a common test for the phenol functional group.

If you have to explain why phenol is more reactive in electrophilic substitution reactions than benzene, your answer must be in terms of the increased electron density of the ring. A common misconception is to imagine that the hydroxyl group stabilises the intermediate ion in the electrophilic substitution mechanism. This is *not* an alternative answer.

Now try this

Explain why phenol is acidic in aqueous solution, whereas ethanol is not. **(3 marks)**

Directing effects in benzene

Groups attached to a benzene ring where an electrophile attacks the ring.

The 2- and 4-directing effect of electron donating groups

When phenol reacts with dilute nitric acid, substitution occurs at the 2 position (next to the carbon attached to the hydroxyl group) and the 4 position (directly opposite the hydroxyl group) on the ring, Substitution does *not* take place at the 3 and 5 positions.

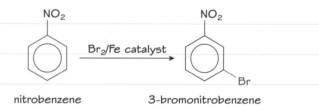

The OH group on the ring is electron donating. It directs the attack to the 2 and 4 positions, shown by the red arrows.

When phenylamine reacts with bromine, the $-NH_2$ group has a similar effect, forming 2,4,6-tribromophenylamine. The $-NH_2$ is also electron donating so activates the ring and directs the attack to the 2 and 4 positions.

Although carbon 6 has undergone substitution, this is still as a result of '2-directing'. This is because the '2-directing' effect applies to both carbon atoms adjacent to the $-NH_2$ group.

The 3-directing effect of electron withdrawing groups

Groups such as $-NO_2$ are electron withdrawing when attached to a benzene ring. This has the effect of deactivating the ring and so slowing down electrophilic substitution reactions. They also direct substitution to the 3 position.

Reaction of nitrobenzene with bromine

NO₂ → (Br₂/Fe catalyst) → NO₂ with Br

nitrobenzene   3-bromonitrobenzene

(a) The aldehyde group in benzaldehyde (C_6H_5CHO) is electron withdrawing. Predict the name of the most likely product when it reacts with chlorine and explain your answer. **(2 marks)**

The most likely product is 3-chlorobenzaldehyde, as electron withdrawing groups are 3-directing.

(b) The $-OCH_3$ group in methoxybenzene ($C_6H_5OCH_3$) is electron donating. Predict its reactivity with bromine in comparison to benzene, give the formula of the likely product and explain your answer. **(3 marks)**

As the $-OCH_3$ group is electron-donating, it will activate the ring meaning it's likely to react faster than benzene.

Electron donating groups are 2- and 4-directing. As this reaction is not required as recall from the specification, any product with substitution at the 2 or 4 positions would be acceptable, for example:

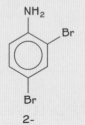

2-bromomethoxybenzene

4-bromomethoxybenzene

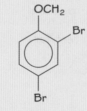

2,4-dibromomethoxybenzene

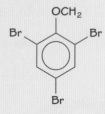

2,4,6-tribromomethoxybenzene

You need to recall the directing effects of $-OH$, $-NH_2$ and $-NO_2$. Other groups may be used in exam questions but you'll be given any necessary information to help you predict products.

Exam skills 11

This exam-style question uses knowledge and skills you have already revised. Look at page 7 and pages 48–54 for a reminder of arenes and the chemistry of benzene and phenol.

Worked example

A student carries out some experiments to compare the reactions of arenes, using methylbenzene and methoxybenzene.

methylbenzene   methoxybenzene

A small amount of bromine solution was added to each arene. The observations noted were:

Arene	Observations when added to bromine
methylbenzene	No visible change
methoxybenzene	Bromine decolourises and misty fumes form. The misty fumes form a white smoke when exposed to ammonia

(a) (i) Suggest a suitable solvent for the bromine used in this experiment and explain your choice. **(3 marks)**

Cyclohexane, as this will mix with both the arenes and the bromine. This is because all three liquids are non-polar, so the only intermolecular forces present are London forces.

(ii) Draw and name the mechanism for the reaction when Br_2 reacts with methoxybenzene. **(4 marks)**

The name of the mechanism is electrophilic substitution.

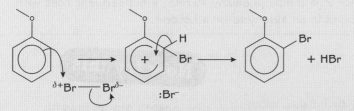

(iii) Explain the difference in the reactivity of the two arenes with bromine. **(3 marks)**

The lone pair of electrons on the oxygen attached to the ring in methoxybenzene is drawn into the delocalised system in the benzene ring. This increases the electron density of the ring, which can then polarise the bromine molecule, making methoxybenzene more reactive than methylbenzene in electrophilic substitution reactions.

As benzene is extremely toxic it is banned from use in schools.

🧪 **Practical skills** Being able to make deductions from qualitative observations such as these, forms a key part of your skills development.

Other hydrocarbon solvents such as hexane will also dissolve both bromine and arenes, so such an exam question would have a range of acceptable solvents.

A more extended answer may go on to compare the relative sizes of the intermolecular forces that exist in the solvent and solute. For instance, the London forces between Br_2 molecules will be similar in size to those between cyclohexane molecules, hence they mix.

Note that as well as showing all curly arrows clearly, all appropriate charges, partial charges and lone pairs are shown.

Methoxybenzene behaves more like phenol than benzene in this reaction, so a halogen carrier such as $FeBr_3$ is not required.

Now try this

The ester group attached to the benzene ring in phenylethanoate will activate the ring. Suggest what effect this will have on the rate of its electrophilic substitution reactions and the orientation of its likely products. **(2 marks)**

Aldehydes and ketones

Aldehydes and ketones are known as carbonyl compounds, as they both contain a carbon atom joined to an oxygen atom with a double bond. Their chemistry is similar, though with some differences, such as their oxidation reactions, explained by the position of the carbonyl group in the molecule.

Aldehydes

Aldehydes have the carbonyl functional group at the end of a carbon chain. For instance propanal is CH_3CH_2CHO.

 — The carbon in the carbonyl group, C=O, is attached to one other carbon atom and a hydrogen atom

Ketones

Ketones have the carbonyl functional group within the carbon chain. For instance propanone is CH_3COCH_3.

 — The carbon in the carbonyl group, C=O, is attached to two other carbon atoms

---

🧪 Practical skills Reactions of carbonyl compounds

As both aldehydes and ketones contain the same functional group, their reactions are very similar. Some notable differences help chemists distinguish between them.

Reactant	Aldehydes	Ketones
$Cr_2O_7^{2-}$ / H^+ (oxidation)	Form a carboxylic acid, colour change from orange to green	No reaction
$NaBH_4$ (reduction)	Form a primary alcohol	Form a secondary alcohol
NaCN / H^+ (nucleophilic addition)	Form a hydroxynitrile	Form a hydroxynitrile
2,4-dinitrophenylhydrazine (nucleophilic addition)	Form a yellow/orange ppt	Form a yellow/orange ppt
Tollens' reagent (oxidation)	Form a silver mirror	No reaction

The reaction with 2,4-dinitrophenylhydrazine can be used to confirm the identity of carbonyl compounds. The yellow/orange precipitate is recrystallised. Once dry, its melting point is found and compared to literature values. The melting point of this derivative is unique so it can be used to confirm the formula of the original aldehyde or ketone.

Tollens' reagent is a mild oxidising agent so it will react with aldehydes but not ketones. Hence if a compound forms a yellow/orange precipitate with 2,4-dinitrophenylhydrazine, a subsequent test with Tollens' reagent will confirm whether the compound is an aldehyde or a ketone.

Worked example

(a) Three unlabelled reagent bottles are known to contain butanal, butan-2-one and butanoic acid. Describe chemical tests which could determine which compound is in which reagent bottle. **(5 marks)**

The three tests used are 2,4-dinitrophenylhydrazine to identify carbonyl compounds; sodium carbonate solution to identify carboxylic acids and Tollens' reagent to distinguish between aldehydes and ketones.

When all three are tested with sodium carbonate solution only the butanoic acid will fizz. The remaining two are then tested with Tollens' reagent. Only butanal forms a silver mirror, so the remaining liquid must be butan-2-one.

(b) Suggest a reason why an old bottle of butanal fizzed slightly when sodium carbonate solution was added. **(2 marks)**

Over time the butanal must have oxidised in air, forming propanoic acid. This reacts with sodium carbonate solution to form carbon dioxide gas.

Now try this

Suggest why testing with Tollens' reagent alone will not confirm whether an unknown compound is an aldehyde or a ketone. **(2 marks)**

Nucleophilic addition reactions

The polar nature of the carbonyl bond means the slightly positive carbon atom is susceptible to attack by electron-rich nucleophiles.

Reaction of carbonyl compounds with NaBH$_4$

The NaBH$_4$ contains the tetrahydridoborate ion, BH$_4^-$. This is a source of the hydride ion, H$^-$, which is the nucleophile in this addition reaction. Nucleophiles are species that are able to donate an electron pair to form a bond in a reaction.

Look at how ethanal reacts with NaBH$_4$ in aqueous solution.

✓ The hydride ions attack the δ$^+$ carbon in the carbonyl bond. The non-bonding pair of electrons of the hydride ion forms a bond to that carbon.

✓ A pair of electrons are transferred from the double bond onto the oxygen, forming an intermediate.

✓ The intermediate then accepts a hydrogen ion from water, which is present as a solvent.

This reaction can also be classified as a reduction, as the aldehyde or ketone gains hydrogen to form an alcohol. Ketones react in a similar way to form a secondary alcohol.

Reaction of carbonyl compounds with HCN

This is a useful reaction in organic synthesis as it extends the length of the carbon chain. The mechanism is analogous to the reaction with the hydride ion. The key difference is the nucleophile – in this case it is the cyanide ion.

Worked example

Draw the mechanism for the reaction of butan-2-one with HCN using skeletal formulae.   **(4 marks)**

Although it's unlikely that you will be specifically asked to use skeletal formulae in a mechanism, it's good practice as formulae in the exam will often be skeletal.

Be careful! Lots of students assume that the non-bonding pair of electrons (or lone pair) on the cyanide ion is found on the nitrogen, perhaps based on previous experience of nucleophiles such as ammonia. However, the pair is actually on the carbon atom so watch out – examiners may be looking for this!

Now try this

(a) Draw a dot-and-cross diagram of the cyanide ion, CN$^-$.   **(2 marks)**

(b) Draw the mechanism for the reaction between cyclohexanone and CN$^-$.   **(3 marks)**

Carboxylic acids

Carboxylic acids are organic compounds containing the -COOH functional group.

Structure of carboxylic acids

The –COOH functional group, present in all carboxylic acids, is called the carboxyl group.

Carboxylic acids are named by taking the alkane stem, removing the –e from the end and adding the suffix 'oic acid'. Hence the molecule above, with a 2-carbon longest chain, is called ethanoic acid.

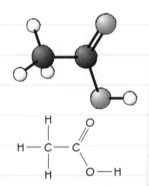

Be careful! Students often see the C=O bond in carboxylic acids and assume it will form a yellow/orange precipitate with 2,4-dinitrophenylhydrazine. It doesn't! This is because the electrons from the carbonyl bond are partly delocalised over the whole carboxyl group. This makes the slightly positive carbon less attractive to nucleophiles.

Solubility in water

Smaller carboxylic acids readily dissolve in water due to the formation of hydrogen bonds such as the ones below.

However, as the length of the non-polar hydrocarbon chain increases, the solubility decreases. For instance, the solubility of pentanoic acid is 4.97 g per 100 cm³ of water, whereas octanoic acid only has a solubility of 0.068 g per 100 cm³ of water.

Reactions with inorganic bases and metals

With inorganic bases and metals, carboxylic acid reacts in the same way as mineral acids, forming carboxylate salts.

Notice in the salt no bond is drawn between the metal and the oxygen, as this part of the molecule is ionic **not** covalent.

Reaction with metals

$$H_3C-C(OH)(=O) + Na \longrightarrow H_3C-C(O^-Na^+)(=O) + H_2$$

Reaction with metal oxides or alkalis

$$H_3C-C(OH)(=O) + NaOH \longrightarrow H_3C-C(O^-Na^+)(=O) + H_2O$$

Reaction with metal carbonates

$$2H_3C-C(OH)(=O) + Na_2CO_3 \longrightarrow 2H_3C-C(O^-Na^+)(=O) + H_2O + CO_2$$

Worked example

Suggest why the boiling point of ethanoic acid is greater than that of propan-1-ol.   **(3 marks)**

Both molecules have the same molar mass (60), so they have the same number of electrons. This implies that their London forces will be similar. Both can also form hydrogen bonds. However, the hydrogen bonds in ethanoic acid can form a **dimer**, where two molecules join together.

This effectively gives the dimer twice the number of electrons, so its London forces will be greater, so it has a higher boiling point.

Now try this

Write equations for the reactions of propanoic acid with:
(a) calcium carbonate   **(2 marks)**
(b) copper oxide.   **(2 marks)**

Esters

Organic esters are compounds where the –OH group in a carboxylic acid is replaced by the –OR group, where R is an alkyl group. Esters with a low relative molecular mass tend to smell fruity and are often used in fragrances.

Formation of esters

The ester butyl ethanoate can be made by the reaction of ethanoic acid with butan-1-ol, with a concentrated sulfuric acid catalyst.

ethanoic acid   butan-1-ol   butyl ethanoate   water

The –OH group from the acid and a hydrogen atom from the alcohol combine to form the other product, water.

Even with a catalyst, this reaction is quite slow and gives a poor yield. A better yield can be obtained by using an acid anhydride with an alcohol.

The ester methyl ethanoate can be formed from methanol and ethanoic anhydride.

ethanoic anhydride   methanol   methyl ethanoate   ethanoic acid

When naming esters, the first part of the name is derived from the alcohol, using the appropriate alkyl stem, in this case methyl. The second part of the name comes from the acid or acid anhydride. The suffix 'oic' is replaced by 'oate', in this case ethanoate.

Hydrolysis of esters

In hydrolysis reactions the esters react with water and break apart. The reaction can be done by heating under reflux, using either acid or alkali as a catalyst. Using H_2SO_4 as a catalyst, methyl benzoate can be hydrolysed to form methanol and benzoic acid.

Using an alkali catalyst, the salt of the acid will form. Adding excess HCl will then re-form the acid.

This bond is broken in the reaction

methyl benzoate   water   benzoic acid   methanol

Worked example

Draw the formula of the two products formed when the ester below is hydrolysed using sodium hydroxide solution. **(2 marks)**

Remember: as this is alkaline hydrolysis, a sodium salt is formed *not* an acid.

Be careful! It's commonplace in exam questions to be asked to apply simple chemistry to quite complex molecules that you've probably never seen before. This is the case here. You just have to hold your nerve and trust the chemistry!

Now try this

Draw the skeletal formula of the ester pentyl hexanoate. Name the *two* products formed when it is hydrolysed under acidic conditions. **(3 marks)**

Acyl chlorides

Acyl chlorides are derived from carboxylic acids, with the hydroxyl group of the acid replaced by a chlorine atom.

The acyl chloride functional group

Acyl chlorides contain the –COCl functional group, for example ethanoyl chloride.

The carbon attached to the oxygen and chlorine atoms is readily attacked by nucleophiles. This is because both the oxygen and chlorine are highly electronegative so withdraw electrons from this carbon. Hence acyl chlorides are more reactive than the acid they are derived from.

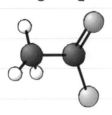

Preparation of acyl chlorides

Acyl chlorides can be prepared by reaction of a carboxylic acid with a chlorinating agent. A suitable chlorinating agent is $SOCl_2$.

For example, ethanoic acid can be converted to ethanoyl chloride:

$$CH_3COOH(l) + SOCl_2(l)$$
$$\rightarrow CH_3COCl(l) + HCl(g) + SO_2(g)$$

> $SOCl_2$ is the preferred chlorinating agent because both waste products are gases so they are easy to separate from the reaction mixture.

Reactions of acyl chlorides

Acyl chlorides such as ethanoyl chloride are used in synthesis to produce esters, carboxylic acids, primary amides and secondary amides.

> In esterification reactions involving phenols, acyl chlorides are often used instead of carboxylic acids. This is because the reaction between a phenol and a carboxylic acid is extremely slow.

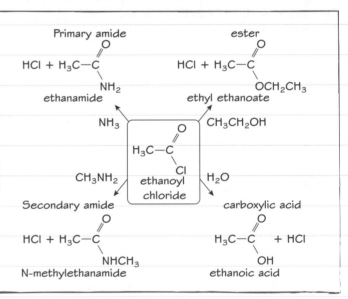

Amides

You're not expected to know much about the reactions of amides but you are expected to recognise them!

Primary amides have the formula $RCONH_2$.

Secondary amides have the formula $RCONHR$.

In both cases, R is an alkyl group.

Worked example

When a bottle of ethanoyl chloride is opened misty fumes are immediately seen. Explain this observation.   **(2 marks)**

Misty fumes are normally a sign that a hydrogen halide has been produced, in this case HCl. Acyl chlorides are so reactive that they will react immediately if exposed to water vapour, forming a carboxylic acid and HCl(g).

Now try this

Write an equation, using skeletal formulae, for each of these reactions:
(a) methylpropanoyl chloride + ammonia   **(1 mark)**
(b) benzoyl chloride + butan-2-ol.   **(1 mark)**

Amines

Amines contain a nitrogen atom. The lone pair on this atom allows them to behave as nucleophiles and as bases.

The amine functional group

Amines such as ethylamine are derived from ammonia, with one or more of the hydrogen atoms replaced by alkyl groups.

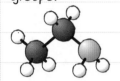

H—C—C—N with H atoms (structural formula of ethylamine)

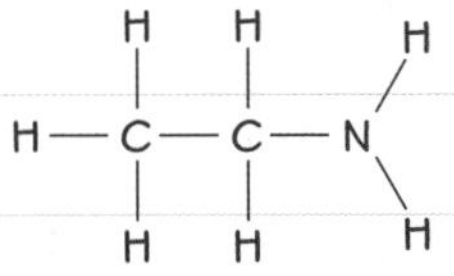

Amines are classified as primary, secondary or tertiary depending on their structure.

ethylamine	N-methylethylamine	N,N-dimethylethylamine
⌒NH₂	⌒NH₂	⌒N⌒
1° as 1 hydrogen replaced by alkyl group	2° as 2 hydrogens replaced by alkyl groups	3° as 3 hydrogens replaced by alkyl groups

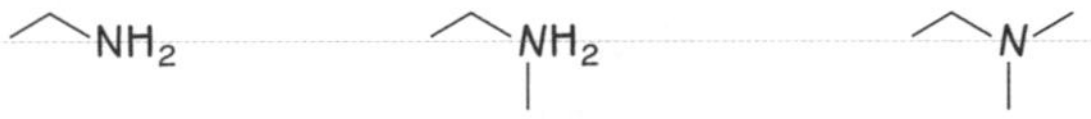

Preparation of amines

- Primary amines can be formed by the reaction of a haloalkane with ammonia dissolved in ethanol. For instance, to form ethylamine you could use brom-ethane and ammonia.

$CH_3CH_2Br + 2NH_3 \rightarrow CH_3CH_2NH_2 + NH_4Br$

They can also be formed by the reduction of nitriles using a reducing agent such as H_2 / Ni.

For instance, to form ethylamine you would use CH_3CN, ethanenitrile.

$CH_3CN + 4[H] \rightarrow CH_3CH_2NH_2$

- Secondary amines can be formed by the reaction of a haloalkane with a primary amine. For instance, to form N-ethylethylamine you could use ethylamine and bromoethane.

$CH_3CH_2NH_2 + CH_3CH_2Br \rightarrow (CH_3CH_2)_2NH + HBr$

- The reaction can continue to form the tertiary amine, N,N-diethylethylamine, as the secondary amine formed reacts with the haloalkane.

$(CH_3CH_2)_2NH + CH_3CH_2Br \rightarrow (CH_3CH_2)_3N + HBr$

- Aromatic amines are formed by reduction of nitroarenes, using tin and concentrated hydrochloric acid as the reducing agent. For instance

$C_6H_5NO_2 + 6[H] \rightarrow C_6H_5NH_2 + 2H_2O$

nitrobenzene　　　　phenylamine

Reactions of amines

Amines behave as bases as the lone pair of electrons on the nitrogen atom can be donated to an electron deficient species such as a hydrogen ion (proton). This forms a dative covalent bond.

Methylamine　　　Proton　　　Methylammonium ion

A salt is formed when the negative ion from the acid, for example a chloride ion, combines with the alkylammonium ion.

$[CH_3NH_3]^+ + Cl^- \rightarrow [CH_3NH_3]^+Cl^-$

　　　　　methylammonium chloride

Had a look ☐ Nearly there ☐ Nailed it! ☐

Amino acids

Amino acids are naturally occurring molecules containing an amine group and a carboxylic acid group.

Structure of amino acids

The 20 naturally occurring amino acids are called α-amino acids and have the general formula $RCH(NH_2)2COOH$. R is an alkyl group, or in the simplest example, glycine, it is another hydrogen atom.

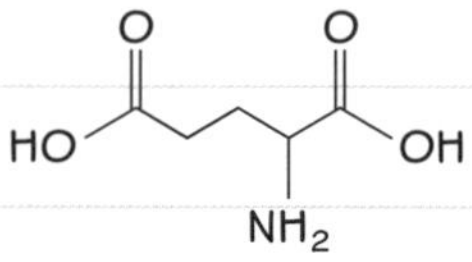

Reactions of carboxylic acid group

The –COOH group reacts with bases in the same way as carboxylic acids.

Reaction with metals

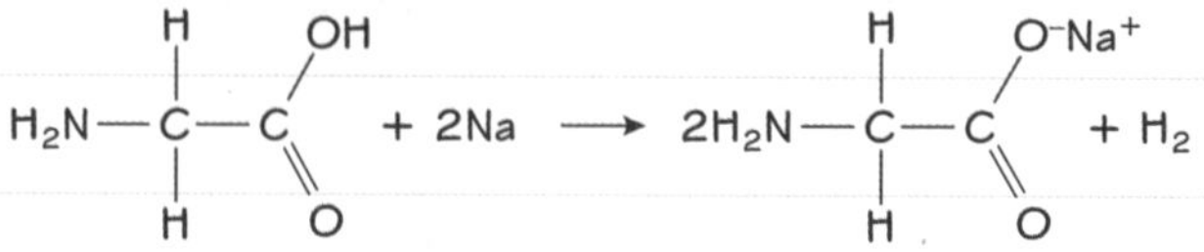

Reaction with metal oxides or alkalis

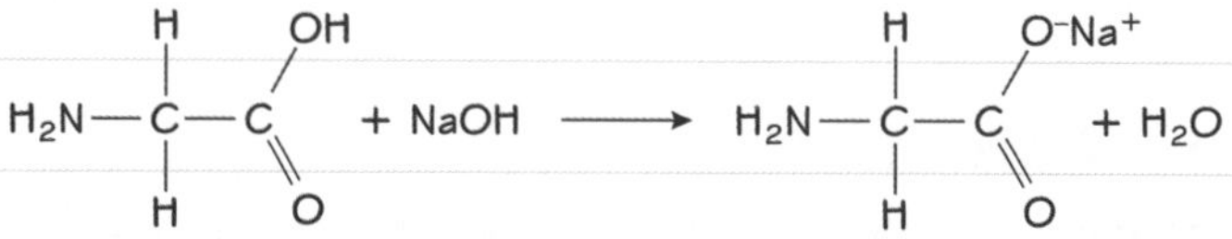

Reaction with metal carbonates

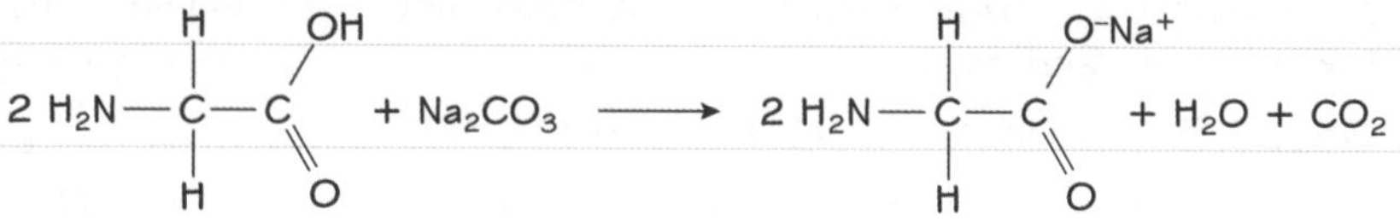

Reactions of amine group

The lone pair on the nitrogen in the amine means amino acids can behave as bases and form salts with acids, e.g. HCl.

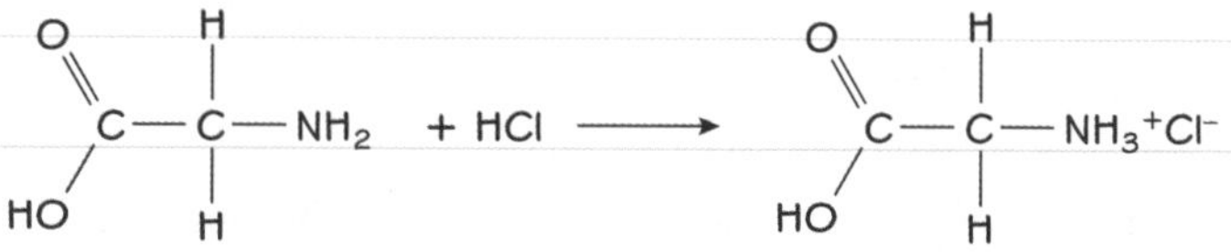

Worked example

Draw the skeletal formula of the organic product formed when the amino acid serine reacts with propan-2-ol, with a hydrochloric acid catalyst. Name the type of organic reaction that takes place.

Serine has the skeletal formula:

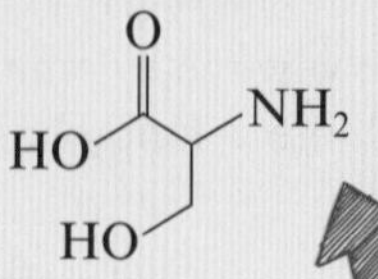

Here you have to be able to interpret and draw a skeletal formula and recognise the clues to help you determine the reaction type. Although serine may be unfamiliar, you should focus on the functional groups.

Under acidic conditions $-NH_2$ reacts as a base with the acid, forming a salt. So, the product formed is:

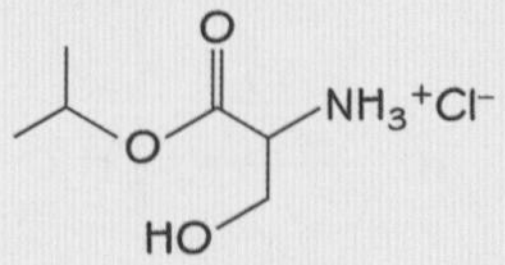

The –COOH group reacts with alcohols to form an ester, so the type of organic reaction is **esterification**.

Now try this

Draw the skeletal formula of the product formed when glutamic acid reacts with ethanol with a hydrochloric acid catalyst. Glutamic acid has the skeletal formula:

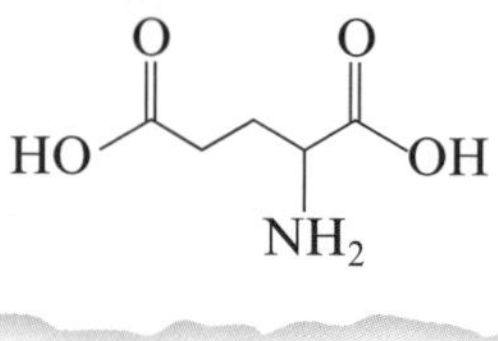

(2 marks)

Optical isomers

Optical isomers are a type of stereoisomer formed when molecules have a chiral centre.

Chiral centres

A molecule has a chiral centre if it contains a carbon atom attached to four different groups. For instance look at the diagram of 1-bromo-1-chloroethane.

The carbon marked with a * has four different groups attached, H, CH_3, Cl and Br. Such molecules have *no* plane of symmetry.

$$H_3C — C^* — Br$$

with Cl above and H below the central carbon.

🖩 Maths skills — Drawing molecules with a chiral centre

It's helpful to draw molecules with chiral centres in a way that emphasises the three-dimensional nature of the molecule.

This bond from the chiral carbon to the methyl group is facing away from you.

Bond in the plane of the paper.

This bond from the chiral carbon to the hydrogen is facing towards you.

Examples of optical isomers

Molecules with a chiral centre have two forms where the atoms have a different arrangement in space. These are optical isomers and are mirror images of each other.

The two isomers, sometimes called enantiomers, are non-superimposable. This means it's impossible to rotate one form to exactly match up with the other.

Chemically, the two forms are virtually identical. The only exceptions are some reactions between molecules with chiral centres in biological systems.

Most synthetic methods of making compounds with chiral centres are **non-stereospecific**, that is they produce a 50:50 mixture of both forms. This is called a **racemic** mixture.

Differences

Polarised light waves are light waves in which the vibrations occur on a single plane.

Each pure sample of an optical isomer will rotate the plane-polarised light to the same degree but in opposite directions.

Worked example

Mark all the chiral centres on the molecule carvone with an asterisk (*).

Explain what effect a racemic mixture of carvone would have on the plane of plane-polarised light. **(3 marks)**

The racemic mixture would have *no effect* on the plane of plane-polarised light. The racemic mixture is a 50:50 mixture of both isomers. The net effect will be for the rotations due to each isomer to cancel each other out.

Now try this

Which of these compounds has a chiral centre?

A $CH_3CH_2C(OH)(Br)CH_2CH_3$

B $CH_3CH_2C(OH)_2CH_2CH_2CH_3$

C $CH_3CH_2C(OH)(Br)CH_3$ **(1 mark)**

Condensation polymers

Polymers such as polyesters are formed by elimination reactions, when monomers join together and release a small molecule such as water.

Polyesters

Polyesters are **condensation** polymers, normally formed when a dicarboxylic acid reacts with a diol, to form a polymer and water.

PET, used in plastic bottles, is an example:

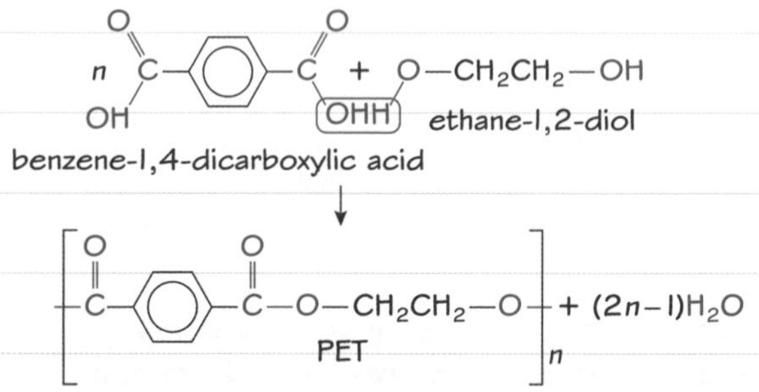

benzene-1,4-dicarboxylic acid   ethane-1,2-diol

PET

Polyamides

Polyamides are **condensation** polymers, normally formed when a dicarboxylic acid reacts with a diamine, to form a polymer and water.

Nylon-6,6, used in synthetic rope, is an example:

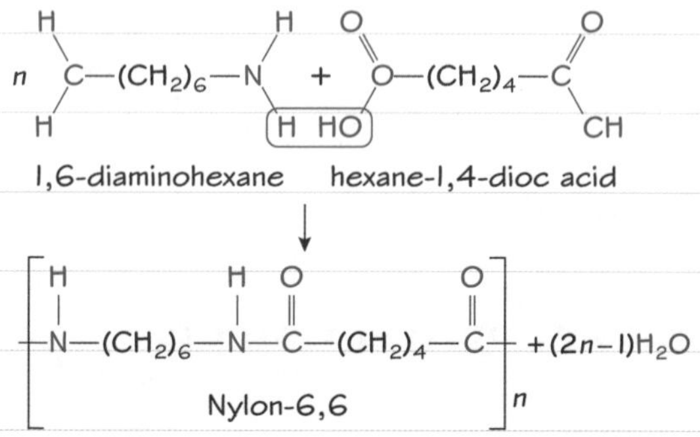

1,6-diaminohexane   hexane-1,4-dioc acid

Nylon-6,6

Hydrolysis of polymers

Condensation polymers can be broken down by the reaction with water – **hydrolysis**. This can be done under both acidic and basic conditions. In polyesters the ester link breaks; in polyamides the amide link breaks. The diagram on the right shows the hydrolysis of PET.

Under basic conditions, the salt of the acid is formed. When polyamides are hydrolysed under acidic conditions the amine groups form a salt.

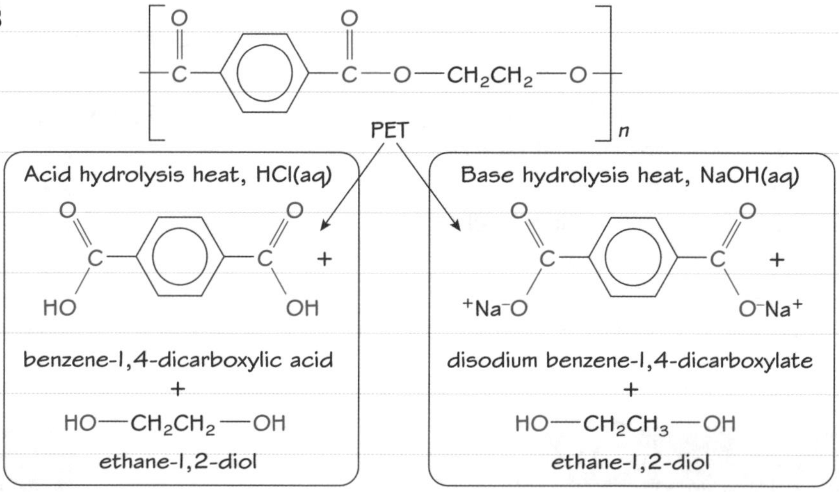

PET

Acid hydrolysis heat, HCl(aq)

benzene-1,4-dicarboxylic acid
+
HO—CH₂CH₂—OH
ethane-1,2-diol

Base hydrolysis heat, NaOH(aq)

disodium benzene-1,4-dicarboxylate
+
HO—CH₂CH₃—OH
ethane-1,2-diol

Draw the products formed when the polymer Qiana undergoes acidic hydrolysis using HCl(aq). **(2 marks)**

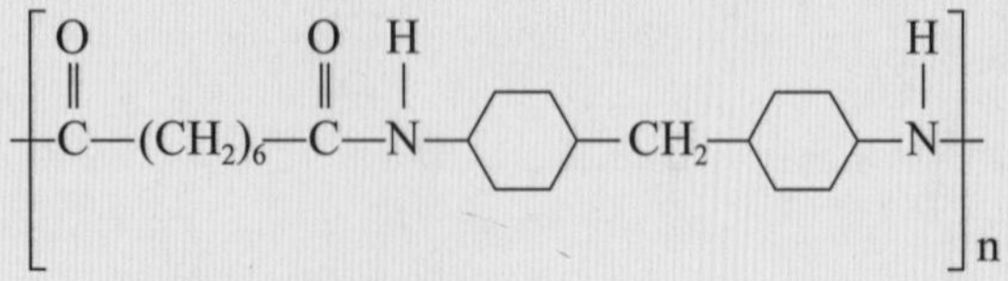

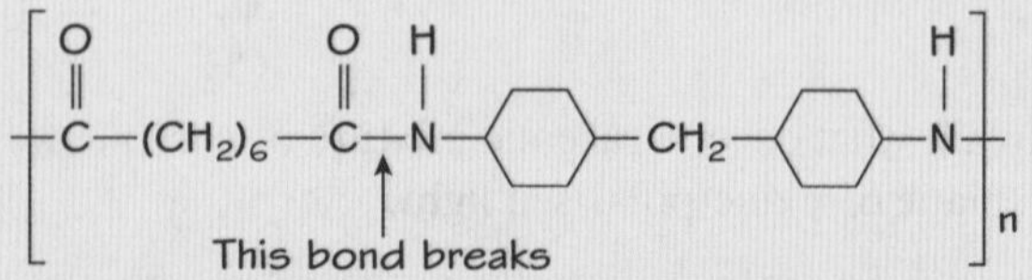

This bond breaks

so the products formed are HOOC(CH₂)₆COOH and

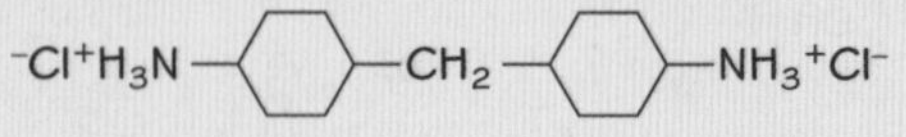

$^-Cl^+H_3N$—◯—CH₂—◯—$NH_3^+Cl^-$

Draw the formula of the repeat unit of the condensation polymer formed from glycolic acid, CH₂(OH)COOH.

(1 mark)

Exam skills 12

This exam-style question uses knowledge and skills you have already revised. Look at pages 133 and 134 for a reminder of optical isomers and the formation and hydrolysis of condensation polymers.

Worked example

Lactic acid can be used to make a biodegradable polymer called poly(lactic acid) or PLA.

(a) Mark the chiral carbon in lactic acid using an asterisk (*). **(1 mark)**

> A *chiral carbon* is a carbon atom attached to four different atoms or groups of atoms. An exam question may ask you to identify several of these within a molecule.

(b) Molecules with a chiral carbon show a particular type of isomerism.

> This type of stereoisomerism is called **optical isomerism** and this would also be an acceptable answer. The two forms are called **enantiomers**.

  (i) Name the type of isomerism shown by lactic acid and show, using 3D diagrams, the two isomeric forms of lactic acid. **(3 marks)**

The type of isomerism shown by lactic acid is stereoisomerism.

> The two isomers are mirror images of each other. If you take one isomer and try to place it over the other, they will not match. They are said to be non-superimposable.

  (ii) Describe the key difference in the properties of the two isomers of lactic acid. **(2 marks)**

> There is no way of telling, from looking at their structure, whether an isomer will rotate the plane of plane-polarised light clockwise or anti-clockwise.

The two isomers will both rotate the plane of plane-polarised light. However one isomer will rotate the light clockwise, the other anti-clockwise.

(c) PLA is a condensation polymer.

> Chemically the two enantiomers will be almost identical. However, in some biochemical reactions only one form will react in a particular way. Such reactions are called **stereospecific**.

  (i) Complete the equation to show the formation of PLA and name the functional group that forms. **(4 marks)**

Ester functional group

> The '2n' used in the question indicated that two repeat units are needed to balance the equation. Only 'n' is used after the repeat unit, as for every 2 monomers, 1 repeat unit of the polymer forms.

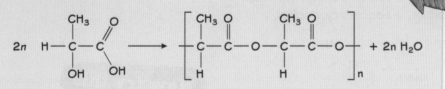

  (ii) PLA can be used to support bones as they mend after a fracture, gradually biodegrading once the bone has healed. Suggest how PLA biodegrades over time. **(2 marks)**

> Hydrolysis of PLA in the laboratory would be much quicker as you could use acidic or basic conditions to speed up the process.

The PLA undergoes a hydrolysis reaction to form lactic acid.
The lactic acid then breaks down further to form carbon dioxide and water.

Carbon–carbon bond formation

Forming carbon-carbon bonds in organic chemistry is useful as it will extend the length of the carbon chain.

Extending the carbon chain in organic compounds

A carbon atom can be added to an aliphatic carbon chain by the formation of nitriles or hydroxynitriles, which can then be converted into other functional groups by reduction or hydrolysis.

Mechanism 1 is nucleophilic substitution (see page 70).

Mechanism 2 is nucleophilic addition (see page 127).

haloalkane
$R-X$

Mechanism 1 ↓

nitrile
$R-CN$

hydrolysis H_2O/H^+ ↙   ↓ reduction H_2/Ni

carboxylic acid
$R-COOH$

amine
$R-CH_2NH_2$

aldehyde or ketone
$$R-\overset{\overset{O}{\parallel}}{C}-R'$$

Mechanism 2 ↓

hydroxynitrile
$$R-\overset{\overset{OH}{|}}{\underset{\underset{CN}{|}}{C}}-R'$$

hydrolysis H_2O/H^+ ↙   ↘ reduction H_2/Ni

hydroxycarboxylic acid
$$R-\overset{\overset{OH}{|}}{\underset{\underset{COOH}{|}}{C}}-R'$$

hydroxyamine
$$R-\overset{\overset{OH}{|}}{\underset{\underset{CH_2NH_2}{|}}{C}}-R'$$

Worked example

(a) Suggest a reaction scheme to convert bromoethane to propan-1-ol, in three steps. Include details of appropriate reagents.   **(4 marks)**

The carbon chain can be extended by forming a nitrile, which can then be hydrolysed to a carboxylic acid. This acid can then be reduced to a primary alcohol.

bromoethane   NaCN/H⁺ →   CN   H₂O/H⁺ →   COOH   NaBH₄ →   OH   propan-1-ol

(b) Suggest a reaction scheme to convert benzene to 1-phenyl ethanol in two steps.   **(4 marks)**

1-phenylethanol

The carbons can be added to the benzene ring by carrying out an acylation reaction to form a phenyl ketone. This can then be reduced to the product, a secondary alcohol.

benzene   AlCl₃ acylation   phenylethanone   NaBH₄ reduction   1-phenylethanol

In both these examples you will have to use your knowledge of organic reactions from across the specification to arrive at your answer. These are examples of simple synthetic routes that specifically require carbon atoms to be added to a chain or ring.

Using skeletal formulae clearly shows how the carbon chain has been extended by one carbon.

Friedal-Crafts Reactions can be used to replace hydrogen atoms by carbon atoms on a benzene ring.

Now try this

Suggest a reaction scheme to form propanoyl chloride from ethanol in four steps. Details of reagents are *not* required.   **(4 marks)**

Purifying organic solids

 **Practical skills** Many organic compounds are crystalline solids. They can be separated and purified by vacuum filtration and recrystallisation. The purity can then be checked by measuring the melting point.

Vacuum filtration

- Vacuum filtration is carried out to collect a solid product.
- A Büchner funnel is used, connected to a side-arm flask made of strong glass, to cope with reduced pressure.
- A circle of filter paper, just slightly smaller than the diameter of the Büchner funnel, is used.
- The reduced pressure speeds up the filtering process and removes more of the solvent. This forms a dryer product.

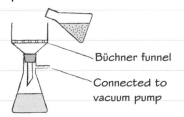

Büchner funnel

Connected to vacuum pump

Recrystallisation

- Add hot solvent dropwise to the impure solid.
- Solid will be sparingly soluble in cold solvent, readily soluble in hot solvent.
- Keep adding solvent until all of the solid just dissolves.
- Allow solution to cool in an ice bath.
- Filter off the solid using vacuum filtration.
- Rinse the solid with a small amount of cold solvent.
- Dry the solid in a desiccator.

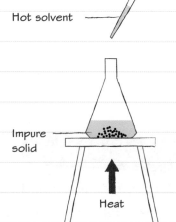

Hot solvent

Impure solid

Heat

Melting point apparatus

Heat gently and record the temperature at which the sample melts. Continue until all the sample melts and record the final temperature.

Place small amount of sample in sealed capillary tube

Thiele tube

Oil

How recrystallisation works

- The hot solvent dissolves both the product and soluble impurities.
- The product is much less soluble in the cold solvent, so forms a precipitate as it cools.
- The impurities remain in solution as they are present in small amounts and more soluble in the chosen solvent.
- A small amount of product will also remain in solution, reducing the yield.
- The solvent used for washing is cold to prevent the product re-dissolving.

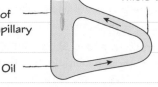 **Worked example**

Describe how to check the purity of your product after recrystallisation. **(3 marks)**

The melting point of the sample can be found using melting point apparatus.

The melting point range is compared to the true value from a reliable data source. A narrow melting point range indicates the sample is relatively pure.

Now try this

Describe how you would remove any insoluble impurities during a recrystallisation. **(2 marks)**

Predicting the properties and reactions of organic compounds

Knowledge of the chemistry of different functional groups and linking together ideas that explain physical properties is used to predict the behaviour of molecules you may not have previously seen.

Paracetamol

The painkiller paracetamol has the structure:

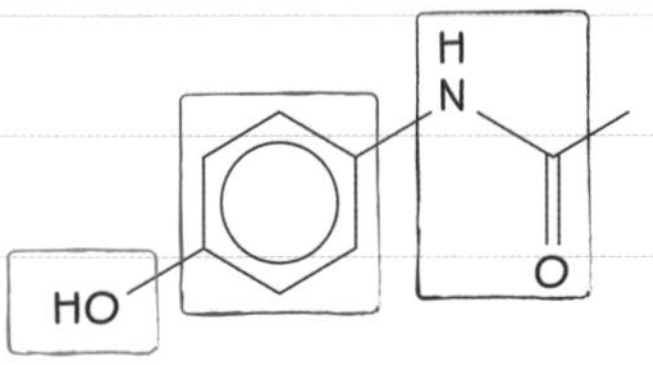

Worked example

1   State the functional groups present in paracetamol   **(3 marks)**

Hydroxyl, arene, amide

The three functional groups are highlighted on the diagram. Although you are not expected to specifically recall the reactions of paracetamol, you could be asked to apply your knowledge of these individual functional groups.

Worked example

2   Suggest why paracetamol is only sparingly soluble in water but readily soluble in ethanol.   **(2 marks)**

There are several groups in paracetamol that can form hydrogen bonds to both water and ethanol, such as the –OH group or the nitrogen in the -NH group. However, the large hydrocarbon benzene ring can form stronger London forces to the ethyl group in ethanol than to water.

3   Draw and name the structures of the products formed when paracetamol reacts with a dilute aqueous solution of sodium hydroxide. **(3 marks)**

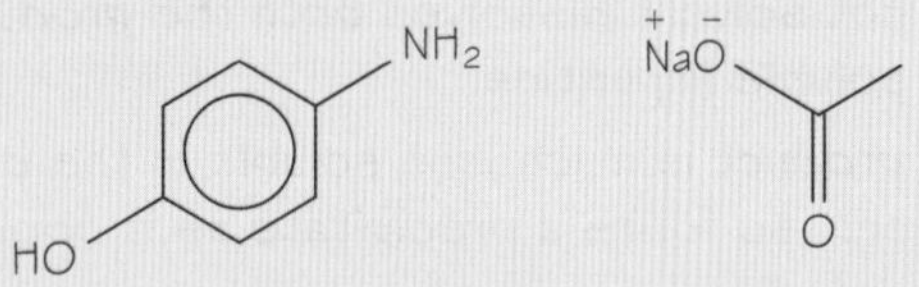

sodium ethanoate

The amide group hydrolyses in a similar way to esters, forming an amine and a carboxylic acid. Note that as the reaction is carried out in alkaline conditions in this case, the salt of the acid forms.

4   Suggest the structure of the product formed when paracetamol reacts with bromine.   **(1 mark)**

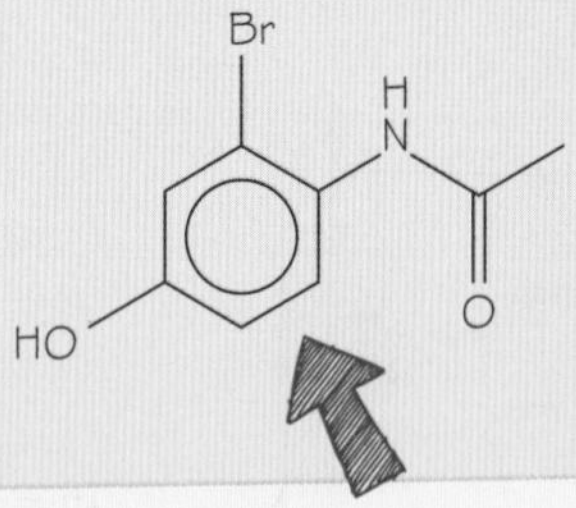

The arene group will undertake electrophilic substitution reactions with bromine in a similar way to phenol.

The structure above shows the presence of only one Br atom, though multiple substitutions may take place and would be an acceptable answer.

Now try this

Suggest the organic products formed when CH_2CHCl:
(a) reacts with HBr   **(1 mark)**
(b) reacts with NaOH(aq)   **(1 mark)**
(c) is heated under pressure with a catalyst.   **(1 mark)**

Summary of organic reactions

Many of the reactions you are expected to recall in organic chemistry are summarised here. You will find it helpful to go through these reactions and produce your own summaries in different formats, relying on looking at this page a little less each time! Remember: although you are expected to know these reactions, the real skill lies in spotting when a reaction is relevant in an unfamiliar compound or situation.

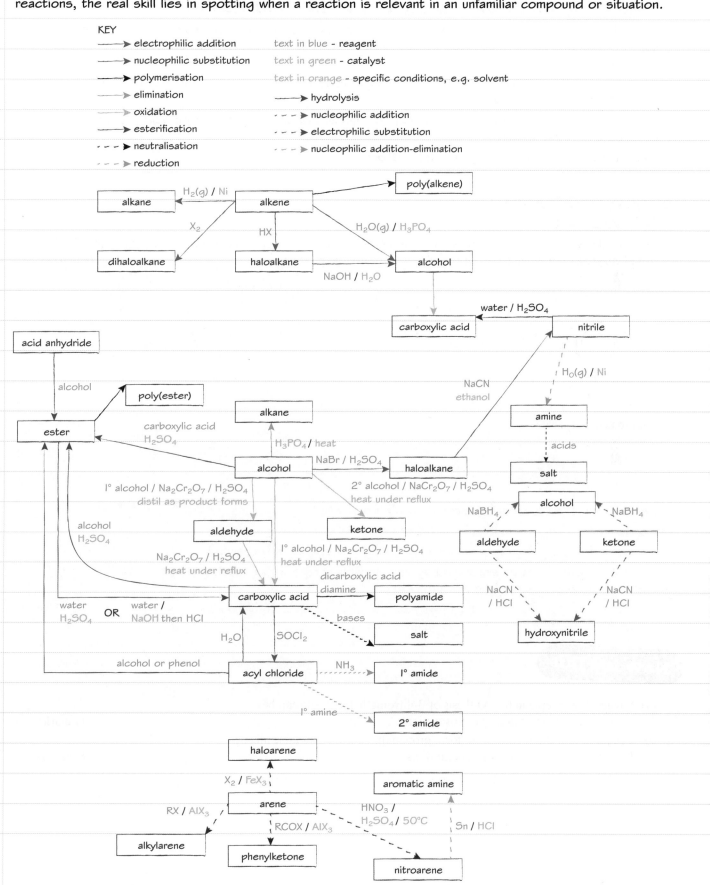

KEY
⟶ electrophilic addition
⟶ nucleophilic substitution
⟶ polymerisation
⟶ elimination
⟶ oxidation
⟶ esterification
- - ➤ neutralisation
- - ➤ reduction

text in blue - reagent
text in green - catalyst
text in orange - specific conditions, e.g. solvent
⟶ hydrolysis
- - ➤ nucleophilic addition
- - ➤ electrophilic substitution
- - ➤ nucleophilic addition-elimination

Organic synthesis

Having studied a number of reactions of different functional groups, a key skill required at A level is to put together sequences of these reactions to convert a simple starting material into a target molecule.

Practical skills

Preparing a sample of 3-nitromethylbenzoate

Step I oxidises the primary alcohol group to a carboxylic acid group. Another suitable oxidising agent is acidified sodium dichromate(VI). The reaction mixture should be heated under reflux.

Step 2 forms the methyl ester, hence the alcohol used should be methanol (CH_3OH). A sulfuric acid catalyst speeds up the reaction.

Step 3 is the nitration of the benzene ring. The mixture of concentrated nitric acid and sulfuric acid is called a nitrating mixture.

The product can be left to dry in a desiccator.

Worked example

Describe the reactions and conditions needed to prepare a sample of 3-nitromethylbenzoate, a crystalline solid, from benzyl alcohol. Include an outline of how to purify the product. **(9 marks)**

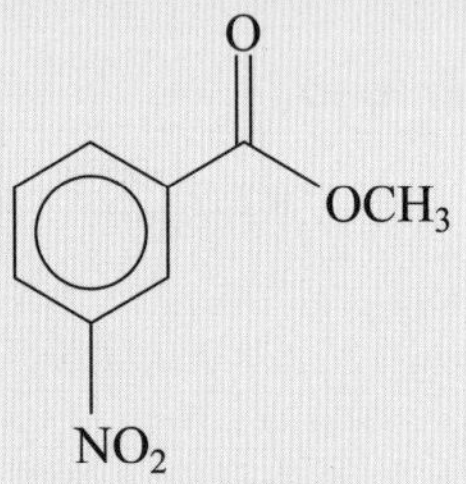

3-nitromethylbenzoate

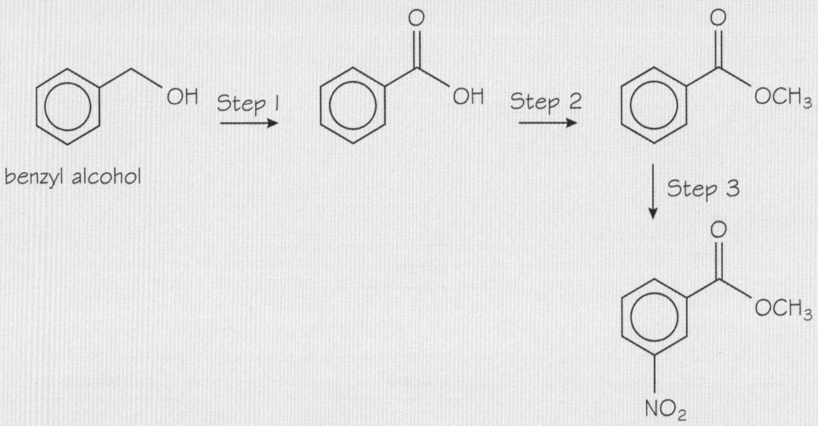

benzyl alcohol → Step I → Step 2 → Step 3 → 3-nitromethylbenzoate

Reagent for Step I – acidified potassium dichromate(VI).

Reagents for Step 2 – methanol (CH_3OH) and sulfuric acid.

Reagents for Step 3 – concentrated nitric acid and concentrated sulfuric acid.

The solid product can be separated by vacuum filtration, recrystallised, washed and then filtered again before leaving to dry.

Now try this

(a) Suggest how Step 3 in the synthesis of 3-nitromethylbenzoate can be controlled to prevent multiple substitutions. **(1 mark)**

(b) Suggest a four-step synthesis to form propanamide from a suitable alkene. Include details of reagents and conditions. **(8 marks)**

Thin layer chromatography

Chromatography can be used to separate and identify different components in a mixture. Thin layer chromatography is used mainly to follow the progress of reactions or assess the purity of products.

Thin layer chromatography

Thin layer chromatography (TLC) is a quick way of testing how far a chemical reaction has gone by checking for the presence of reactants and products. It is also used to test the purity of a product sample.

A spot of the reaction mixture is placed onto the TLC plate – a piece of plastic coated with a stationary phase such as silica. This is placed into a container with a shallow layer of solvent – the mobile phase. The solvent rises up the plate and carries each soluble component of the mixture to different positions on the plate, depending on the difference in strength between the interaction with the stationary phase and with the mobile phase.

An R_f value can be calculated to show how far a component has moved compared to the solvent.

This can be used to identify a compound by comparing to known R_f values.

$$R_f = \frac{\text{distance moved by component}}{\text{distance moved by solvent front}}$$

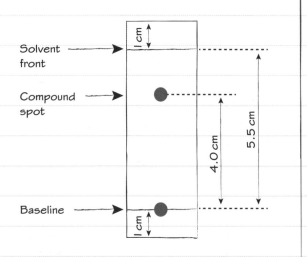

Worked example

(a) Calculate the R_f value in the TLC chromatogram above. **(2 marks)**

$$R_f = \frac{\text{distance moved by component}}{\text{distance moved by solvent front}}$$

$$R_f = 4.0 \div 5.5 = 0.73$$

> Note that as R_f values are ratios of distances, they do not have units.

(b) Describe how TLC could be used to monitor the progress of a chemical reaction. **(2 marks)**

Take samples of the reaction mixture at regular time intervals during the reaction. Run a TLC plate and compare the R_f values to known R_f values or against spots of reactants and expected products on the same TLC plate. The reactant spot disappears and the product spot appears.

Now try this

On a TLC plate, the distance moved by compound X was 4.9 cm. The distance moved by the solvent front was 9.3 cm. Calculate the R_f value for compound X. **(2 marks)**

Gas chromatography

Gas chromatography can be used to separate and identify different volatile components in a mixture. It is often used alongside mass spectrometry.

Gas chromatography

✓ Gas chromatography separates out the components of a mixture of volatile compounds. It uses an inert carrier gas to move the compounds through the stationary phase, a solid compound lining the column.

✓ As different compounds form different strength interactions with the stationary phase, their speeds as they pass through the column to the detector will differ from each other.

✓ The time for a compound to travel from the start of the column to the detector is called the **retention time**. This can be used to suggest the identity of a compound by comparison with chromatograms of known compounds.

✓ The area under a peak in a chromatogram is proportional to the amount of a compound present in the sample.

✓ Calibration curves can be used to find the concentrations of components in a mixture. A chromatogram of a known concentration of a compound is obtained. The area under the peak due to that compound is proportional to its concentration. This can then be compared with the area under peaks in sample chromatograms and the concentration in the sample can then be determined.

Diagram labels: Flow controller, Sample injector, Column, Waste, Carrier gas, Thermostatic oven, Detector

Worked example

Look at the gas chromatogram below then answer the questions that follow.

(a) Which compound has the shortest retention time?

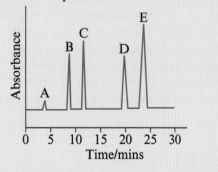

(1 mark)

Compound A has the shortest retention time.

(b) Which compound is present in the greatest quantity? (1 mark)

Compound E has the greatest area under its curve, so it is present in the greatest quantity.

(c) Which compound has the greatest attraction to the stationary phase? (1 mark)

Compound E has the greatest retention time so it must have formed the strongest attraction to the stationary phase.

(d) Explain how the gas chromatogram could be used to suggest the identity of compounds A–E. (2 marks)

By comparing retention times to known values. This can only be an estimate though, because many compounds have very similar retention times.

Now try this

Explain why oxygen would not be a suitable carrier gas in gas chromatography. (2 marks)

Qualitative tests for functional groups (1)

Evidence for the presence of many functional groups can be found by carrying out some simple test-tube reactions. It's important to be able to recall these to help support more complex data from instrumental techniques (e.g. NMR) and so identify compounds.

🧪 Practical skills   Tests for functional groups

Functional group being tested	Details of test	Expected observations	Comments
Alkenes	Gently shake compound with $Br_2(aq)$	Orange to colourless	Do not expose to UV light as alkanes may also give a positive test under these conditions
Haloalkanes	Place compound with $AgNO_3(aq)$ and ethanol in a test tube. Warm in a water bath	White precipitate forms if haloalkane was RCl; cream precipitate forms if haloalkane was RBr; yellow precipitate forms if haloalkane was RI	The white precipitate dissolves in dilute ammonia, the cream precipitate in concentrated ammonia, the yellow precipitate does not dissolve in ammonia, regardless of the concentration
Phenols	Add drops of blue litmus to compound, then shake with small amount of $Na_2CO_3(aq)$	Litmus should turn red, but *no* effervescence would be seen with $Na_2CO_3(aq)$	An alternative test is the formation of a white precipitate with $Br_2(aq)$.
Carboxylic acids	Add drops of blue litmus to solution of compound, then shake with small amount of $Na_2CO_3(aq)$	Litmus should turn red, *and* effervescence would be seen with $Na_2CO_3(aq)$	

Worked example

Describe simple qualitative tests that could be used to distinguish between the compounds below. Include details of the reagents and conditions used.

Write equations for any reactions that take place.

Compound 1   Compound 2   Compound 3
OH   O   O
                 OH   OH

Test all three compounds by placing 1 cm³ of compound with 1 cm³ of Tollens' reagent in a test tube. Warm in a water bath.

Compound 1 will form a silver 'mirror' precipitate as it contains an aldehyde functional group.

$CH_3CH_2CH_2CH(OH)CH_2CH_2CHO + [O]$
$\rightarrow CH_3CH_2CH_2CH(OH)CH_2CH_2CO_2H$

Then test compounds 2 and 3 by placing 1 cm³ of compound with 1 cm³ of acidified dichromate ($Cr_2O_7^{2-}(aq)/H^+$) in a test tube. Warm in a water bath.

Compound 3 will cause a colour change from orange to green, it is a 2° alcohol so can be oxidised to a ketone. Compound 2 cannot be oxidised as it is a 3° alcohol so there will be no colour change when it is tested.

$CH_3CH(OH)CH_2COCH_2(CH_3)CH_3 + [O]$
$\rightarrow CH_3COCH_2COCH_2(CH_3)CH_3 + H_2O$

Now try this

Describe chemical tests to distinguish between these compounds:

OH

COOH

(2 marks)

Qualitative tests for functional groups (2)

Evidence for the presence of many functional groups can be found by carrying out some simple test-tube reactions. It's important to be able to recall these to help support more complex data from instrumental techniques (e.g. NMR) and so identify compounds.

🧪 Practical skills Tests for functional groups

Functional group being tested	Details of test	Expected observations	Comments
Carbonyl compounds (aldehydes or ketones)	Add a few drops of the compound to 1 cm^3 of 2,4-dinitrophenylhydrazine (2,4-DNP)	Orange precipitate forms	Esters and carboxylic acids *do not* react with 2,4-DNP
Aldehydes (often used after a positive 2,4-DNP test to distinguish between aldehydes and ketones)	Place 1 cm^3 of compound with 1 cm^3 of Tollens' reagent in a test tube. Warm in a water bath	Silver 'mirror' precipitate forms	An alternative is to warm with acidified $Cr_2O_7^{2-}$(aq). Aldehydes will react (orange to green colour change), ketones will not
1° and 2° alcohols	Place 1 cm^3 of compound with 1 cm^3 of acidified dichromate ($Cr_2O_7^{2-}$(aq)/ H$^+$) in a test tube. Warm in a water bath	1° and 2° alcohols will react, giving an orange to green colour change	3° alcohols will not be oxidised by acidified dichromate as this involves breaking a very strong C–C bond

Worked example

Test all three compounds by placing 1 cm^3 of compound with 1 cm^3 of Tollens' reagent in a test tube. Warm in a water bath.

Describe simple qualitative tests that could be used to distinguish between the compounds below. Include details of the reagents and conditions used.

Write equations for any reactions that take place.
(6 marks)

Compound 1 will form a silver 'mirror' precipitate as it contains an aldehyde functional group.

$CH_3CH_2CH_2CH(OH)CH_2CH_2CHO + [O]$
$CH_3CH_2CH_2CH(OH)CH_2CH_2CO_2H$

Then test compounds 2 and 3 placing 1 cm^3 of compound with 1 cm^3 of acidified dichromate ($Cr_2O_7^{2-}$(aq)/H$^+$) in a test tube. Warm in a water bath.

Compound 3 will cause a colour change from orange to green, it is a 2° alcohol so can be oxidised to a ketone. Compound 2 cannot be oxidised as it is a 3° alcohol so there will be no colour change when it is tested.

$CH_3CH(OH)CH_2COCH_2(CH_3)CH_3 + [O]$
$CH_3COCH_2COCH_2(CH_3)CH_3 + H_2O$

Now try this

Suggest why carboxylic acids *do not* react with 2,4-dinitriphenylhydrazine.
(2 marks)

Think about how orbital interaction stabilises benzene.

Carbon-13 NMR spectroscopy

Nuclear magnetic resonance (NMR) spectroscopy is the study of how energy in the radio wave section of the electromagnetic spectrum interacts with the nuclei of certain isotopes, for example carbon-13. This gives information that can help determine the structure of a molecule.

C-13 NMR spectrum

A C-13 NMR spectrum is useful as it can determine the number of unique carbon environments present in a molecule. The chemical shift, the point in the spectrum where the nucleus absorbs energy, can also be matched to specific types of carbon environments. Information about the different chemical shifts can be found in data booklets.

The molecule $CH_2=CHCOCH_3$ contains both the alkene and ketone functional groups. The carbons in these groups can be matched to the C-13 NMR spectrum.

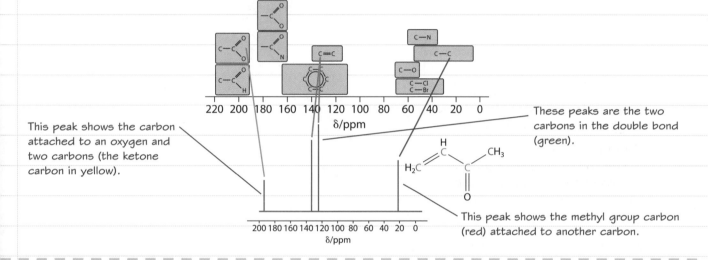

This peak shows the carbon attached to an oxygen and two carbons (the ketone carbon in yellow).

These peaks are the two carbons in the double bond (green).

This peak shows the methyl group carbon (red) attached to another carbon.

Worked example

Explain how C-13 NMR spectra can be used to distinguish between propanal and propanone. Use chemical shift data to predict where peaks would be expected on spectra of both compounds.   **(5 marks)**

propanal   propanone

The easiest way to distinguish between the two compounds is to realise that propanal has 3 unique carbon environments, whereas propanone has two.

In propanone there is one peak around 200 ppm – due to a carbon attached to an oxygen. There is another peak at around 0–50 ppm – this is due to two methyl groups in identical environments where both are attached to another carbon atom.

In propanol there is one peak around 200 ppm – due to a carbon attached to an oxygen – and two peaks around 0–50 ppm. These are due to carbons attached to other carbons.

Now try this

Draw skeletal formulae of 1,2-dimethylbenzene and its three structural isomers.
Predict the number of different peaks in the C-13 NMR spectrum of each isomer. **(4 marks)**

Proton NMR spectroscopy

The nuclei of a hydrogen atom will interact with energy in the radio part of the electromagnetic spectrum. As all organic compounds contain hydrogen atoms, this can provide useful information about the structure of a compound. As the nucleus of the hydrogen atom contains only one proton, the technique is often called proton NMR.

Key features of a proton NMR spectrum

The information about the structure of a compound provided by a proton NMR spectrum includes:

✓ The number of peaks, which indicates the number of unique proton environments.

✓ Chemical shifts, which indicate the type of proton environment.

✓ The ratio of areas under each peak, which indicates the relative number of each type of proton.

✓ Splitting patterns, which is linked to the number of protons attached to the **adjacent** carbon.

H-1 (proton) NMR spectrum of ethanol illustrates these points

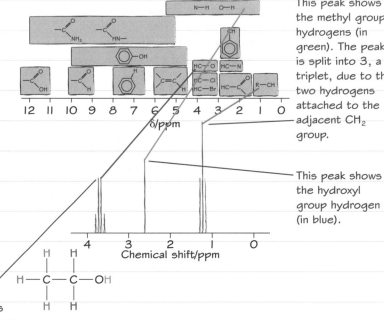

This peak shows the methyl group hydrogens (in green). The peak is split into 3, a triplet, due to the two hydrogens attached to the adjacent CH_2 group.

This peak shows the hydroxyl group hydrogen (in blue).

This peak shows the hydrogens in the CH_2 group (in red). The peak splits into 4, a quartet, as it has 3 neighbouring protons on the adjacent methyl group. The rule for splitting is called the $n+1$ rule. So if a proton has n neighbours on an adjacent carbon, its peak will split into $n+1$ peaks. In this case, 3 neighbouring protons causes splitting into 4 peaks.

Worked example

(a) State the relative areas under each peak in the H-1 NMR spectrum of ethanol. **(3 marks)**

The relative areas under the peaks correspond to the numbers of each type of proton, so the peak at 2.6 ppm will have a relative area of 1 (1 proton in the hydroxyl group).

The peak at 3.7 ppm will have a relative area of 2 (2 protons in the CH_2 group).

The peak at around 1.2 ppm will have a relative area of 3 (3 protons in the CH_3 group).

(b) Predict the splitting pattern in a proton NMR spectrum for a hydrogen that has 6 neighbouring hydrogen atoms on adjacent carbons. Explain your answer. **(2 marks)**

Six neighbouring hydrogen atoms, for instance a CH_3 group on either side, will split the peak into 7 (a septet). This is predicted by the $n+1$ rule; that is, 6 neighbouring hydrogen atoms split the peak into 6 + 1 peaks.

Now try this

Predict the chemical shifts and relative peak areas in the proton NMR spectrum of C_6H_5CHO. **(4 marks)**

Identifying the structure of a compound from a proton (H-1) NMR spectrum

A proton NMR spectrum of an organic compound is used to deduce a possible structure.

Use of TMS in H-1 NMR

The chemical shift in NMR spectroscopy is a way of determining the point at which the nucleus absorbs energy. Absolute values are difficult to determine so the shift is measured relative to the absorption due to the protons in the compound TMS. It has the formula $Si(CH_3)_4$.

TMS is used because:

- It has 12 equivalent protons that give a sharp distinctive single peak, a singlet.
- The peak appears on the far right of the spectrum, well away from other peaks.
- It is non-toxic.
- It is chemically inert.

The position of the peak due to TMS is given the value of 0 ppm.

Worked example

Compound Q is a fruity smelling liquid with the molecular formula $C_4H_8O_2$. Use its H-1 NMR spectrum and data on chemical shifts to suggest a structure and name for compound Q. **(8 marks)**

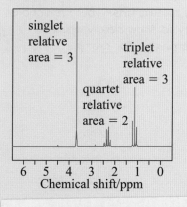

singlet relative area = 3

triplet relative area = 3

quartet relative area = 2

6   5   4   3   2   1   0
Chemical shift/ppm

Q is probably an ester as it smells fruity and has a molecular formula that could match the structure of an ester.

The chemical shifts give information about the environments of the protons, whilst the relative peak areas tell us the number of each proton in a particular environment. See page 146 for chemical shift data.

The peak at about 2.4 ppm corresponds to HCCO.

The relative peak area of 2 suggests a H_2CCO group.

The peak at about 1.2 ppm suggests a R–CH group.

The relative peak area of 3 suggests a $-CH_3$ group.

The splitting of the peak due to the H_2CCO group into a quartet tells us that the $-CH_2$ group is adjacent to a $-CH_3$, as the neighbouring 3 hydrogens cause a 3 + 1 splitting pattern.

The splitting of the peak due to the $-CH_3$ group into a triplet tells us that the $-CH_3$ group is adjacent to a $-CH_2^-$, as the neighbouring 2 hydrogens cause a 2 + 1 splitting pattern.

The structure of compound Q is:

The name of compound Q is methyl propanoate.

Now try this

Describe how the H-1 NMR spectrum of ethyl ethanoate will be similar to and how it will be different from the H-1 NMR structure of methyl propanoate. **(4 marks)**

Predicting a proton NMR spectrum

You may be given the structure of an organic compound and be expected to predict the likely proton NMR spectrum as well as explain the use of different solvents.

Use of solvents in H-1 NMR

Hydrocarbon solvents are not used to dissolve samples for use in H-1 NMR because they contain hydrogen atoms, so the spectra produced would contain peaks due to the solvent as well as the compound.

Solvents containing deuterium, an isotope of hydrogen, are used instead. These do not produce a peak in a NMR spectrum. The most commonly used solvent is $CDCl_3$. The solvent D_2O, deuterium oxide, is also used. This helps identify peaks due to protons in –OH and –NH groups. The deuterium atoms in D_2O exchange places with the protons in an –OH or –NH group over time. A spectrum can be run before D_2O is added. D_2O is then added to the sample and a second spectrum is run. The peak due to the OH or NH will be absent in the second spectrum. By comparing the two spectra this peak can be identified in the original spectrum.

Worked example

(a) Write an equation for the reaction between D_2O, deuterium oxide, and ethanol. **(1 mark)**

$CH_3CH_2OH + D_2O \rightleftharpoons CH_3CH_2OD + HOD$.

(b) Predict the proton NMR spectrum of the compound $CH_3CH_2OCH_2CH_2COCH_3$. Include details of chemical shifts, relative peak areas and splitting patterns. **(5 marks)**

Proton(s) in structure	Chemical shift/ppm	Relative peak area	Splitting pattern
H_3C-C	0.8–2	3	triplet
H_2C-O	3.2–4.2	2	quartet
H_2C-O	3.2–4.2	2	triplet
H_2C-C=O	2–3	2	triplet
H_3C-C=O	2–3	3	singlet

(c) Explain why $CDCl_3$ is often used as a solvent in NMR spectroscopy. **(3 marks)**

It contains no hydrogen atoms. Deuterium atoms do not interact with the radio waves as they have an even number of particles in the nucleus.

It's difficult to draw the spectrum as estimating the relative peak areas is difficult and giving a precise value for each chemical shift is not possible. If you choose to draw the spectrum, clearly label the relative peak area and don't worry if your actual peaks are not to scale. Make sure the chemical shift is within the range given on the data sheet. Another approach is to summarise your predictions in a table, using a key to show clearly which protons you are referring to. For example:

Now try this

Explain the splitting patterns in the predicted spectrum for $CH_3CH_2OCH_2CH_2COCH_3$. **(3 marks)**

Deducing the structure of a compound from a range of data

Candidates are often asked to identify the structure of a compound, explaining how they have arrived at their answer. Information given to help identify the structure can include details of IR and NMR spectra.

Worked example

Compound X is a food additive with a rose-like odour. It has the empirical formula C_5H_6O.
Use this data and the spectra to deduce the structure of X. In each case explain how you used the data.

(8 marks)

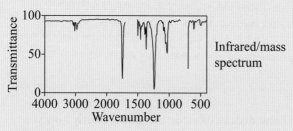
Infrared/mass spectrum

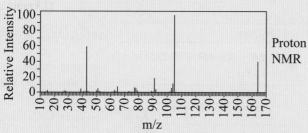
Proton NMR

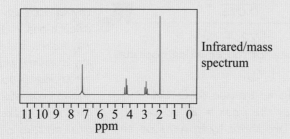
Infrared/mass spectrum

Infrared

The peak in the IR spectrum at about 1730 cm⁻¹ suggests the presence of a C=O group, so X could be an aldehyde, ketone, carboxylic acid or ester. However, the peak at about 1230 cm⁻¹ indicates a C–O group. The lack of a significant broad peak at around 2500–3500 cm⁻¹ suggests that there is no –OH group. Hence X is not likely to be a carboxylic acid or a molecule with both alcohol and carbonyl functional groups. The presence of peaks due to C=O and C–O bonds and the rose-like odour suggests an ester.

Mass spectrum

The m/z value of the molecular ion peak is 164. This is twice the mass of the empirical formula (82) so confirms the molecular formula is $C_{10}H_{12}O_2$.

The large peak at $m/z = 1054$ is the most significant fragment. One possible ion that could have caused this is $C_6H_5CH_2CH_2^+$.

The peak at $m/z = 43$ could be due to a CH_3CO^+ ion.

Proton NMR

The complex peak in the NMR spectrum at 7–7.5 ppm is due to protons attached to a benzene ring.

The triplet at just over 4 ppm is due to a HC–O group. The splitting into a triplet suggests a –CH₂ group is adjacent. The peak at just below 3 ppm is the adjacent –CH₂ group, as it also splits into a triplet. Its chemical shift suggests it is due to hydrogens attached to a carbon next to a benzene ring.

The singlet at 2 ppm is due to HCC=O protons with no neighbouring protons, as no splitting occurs.

Hence the structure of compound X is:

Now try this

State how many peaks you would see in the C-13 NMR spectrum of compound X. **(1 mark)**

Exam skills 13

This exam-style question uses knowledge and skills you have already revised. Look at page 7 and pages 146–148 for a reminder of empirical formulae and proton NMR spectroscopy.

Worked example

Compound G was investigated using combustion analysis, mass spectrometry and proton NMR spectroscopy. The data collected is summarised below.

Combustion analysis Complete combustion of 5.24 g of compound G produced 12.8 g of CO_2 and 5.24 g of H_2O.
Mass spectrometry m/z ratio for the molecular ion = 144
Proton NMR

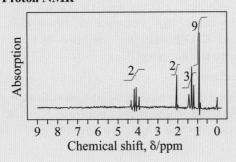

In extended questions such as these don't be afraid to put your answer into sections to help make sure you cover all aspects of the question. In this example, the question is in three sections: combustion analysis, mass spectrometry, and NMR. Try to give your answer in the same way. NMR can be sub-divided further to allow you to show clearly how you used the chemical shifts, splitting patterns and the area under each peak.

Use the data to confirm the molecular formula of compound G and deduce its structure. Show clearly how you used the data to arrive at your answers. **(13 marks)**

Element	Carbon	Hydrogen	Oxygen
Mass	$12.8 \times \left(\dfrac{12}{44}\right) = 3.49\,g$	$5.24 \times \left(\dfrac{2}{18}\right) = 0.58\ g$	$5.24 - (3.49 + 0.58)$ $= 1.17\,g$
Amount	$\dfrac{3.49}{12.0} = 0.29\ mol$	$\dfrac{0.58}{1.0} = 0.58\ mol$	$\dfrac{1.17}{16.0} = 0.073\ mol$
Ratio	$\dfrac{0.29}{0.073} = 4$	$\dfrac{0.58}{0.073} = 8$	$\dfrac{0.073}{0.073} = 1$

Maths skills To help simplify the ratio divide each amount by the smallest amount, in this case 0.073 mol.

Empirical formula = C_4H_8O, which has a relative mass of 72. $\frac{m}{z}$ ratio for the molecular ion = 144, which is double the relative mass of the empirical formula.
So the molecular formula of compound G must be $C_8H_{16}O_2$.

Laying out such calculations in this way makes it clear to any examiner what you are trying to do. This shows understanding even if you make a small mistake in the calculation.

The peak in the NMR spectrum at δ ≈ 4 ppm is likely to be due to O–CH protons and there are two of them as the area under the peak = 2, so O–**CH₂**-. The quartet splitting pattern suggests there are three hydrogens on an adjacent carbon, so O–**CH₂CH₃**.

The peak at δ ≈ 1.2–1.5 ppm supports this, as it corresponds to R–CH protons and there are three of them (area under peak is 3). It also splits into a triplet so the adjacent carbon must have two hydrogen atoms.

Remember to refer to the data sheet to find information on chemical shifts.

The peak at δ ≈ 2 ppm is likely to be due to O=CCH protons and the area under the peak shows there is two such protons.
The peak is a singlet so the adjacent carbon has no hydrogens attached. So the structure so far could be:

$$H_2C - C - O - \overset{H_2}{C} - CH_3$$
$$\quad\ \ \| $$
$$\quad\ \ O$$

The peak at δ ≈ 0.9 ppm also corresponds to R–CH protons and there are 9 of them (area under peak = 9). As the peak is a singlet the adjacent carbon has no hydrogens attached.

Hence this group could be:

$$H_3C - \overset{\displaystyle |}{\underset{\displaystyle |}{C}} - CH_3$$
$$CH_3$$

So overall the structure is:

$$H_3C - \overset{CH_3}{\underset{CH_3}{C}} - \overset{H_2}{C} - C - O - \overset{H_2}{C} - CH_3$$
$$\qquad\qquad\qquad\ \| $$
$$\qquad\qquad\qquad\ O$$

Periodic table

Key

| Atomic number |
| Atomic symbol |
| Name |
| Relative atomic mass |

Example:

| 1 |
| **H** |
| Hydrogen |
| 1.0 |

Group

Period	1 (1)	2 (2)	(3)	(4)	(5)	(6)	(7)	(8)	(9)	(10)	(11)	(12)	3 (13)	4 (14)	5 (15)	6 (16)	7 (17)	8 (18)
1	1 **H** Hydrogen 1.0																	2 **He** Helium 4.0
2	3 **Li** Lithium 6.9	4 **Be** Beryllium 9.0											4 **B** Boron 10.8	6 **C** Carbon 12.0	7 **N** Nitrogen 14.0	8 **O** Oxygen 16.0	9 **F** Fluorine 19.0	10 **Ne** Neon 20.2
3	11 **Na** Sodium 23.0	12 **Mg** Magnesium 24.3											13 **Al** Aluminium 27.0	14 **Si** Silicon 28.1	15 **P** Phosphorus 31.0	16 **S** Sulfur 32.1	17 **Cl** Chlorine 35.5	18 **Ar** Argon 39.9
4	19 **K** Potassium 39.1	20 **Ca** Calcium 40.1	21 **Sc** Scandium 45.0	22 **Ti** Titanium 47.9	23 **V** Vanadium 50.9	24 **Cr** Chromium 52.0	25 **Mn** Manganese 54.9	26 **Fe** Iron 55.8	26 **Co** Cobalt 58.9	28 **Ni** Nickel 58.7	29 **Cu** Copper 63.5	30 **Zn** Zinc 65.4	31 **Ga** Gallium 69.7	32 **Ge** Germanium 72.6	33 **As** Arsenic 74.9	34 **Se** Selenium 79.0	35 **Br** Bromine 79.9	36 **Kr** Krypton 83.8
5	37 **Rb** Rubidium 85.5	38 **Sr** Strontium 87.6	39 **Y** Yttrium 88.9	40 **Zr** Zirconium 91.2	41 **Nb** Niobium 92.9	42 **Mo** Molybdenum 95.9	43 **Tc** Technetium (98)	44 **Ru** Ruthenium 101.1	45 **Rh** Rhodium 102.9	46 **Pd** Palladium 106.4	47 **Ag** Silver 107.9	48 **Cd** Cadmium 112.4	49 **In** Indium 114.8	50 **Sn** Tin 118.7	51 **Sb** Antimony 121.8	52 **Te** Tellurium 127.6	53 **I** Iodine 126.9	54 **Xe** Xenon 131.3
6	55 **Cs** Caesium 132.9	56 **Ba** Barium 137.3	57 **La*** Lanthanum 138.9	72 **Hf** Hafnium 178.5	73 **Ta** Tantalum 180.9	74 **W** Tungsten 183.8	75 **Re** Rhenium 186.2	76 **Os** Osmium 190.2	77 **Ir** Iridium 192.2	78 **Pt** Platinum 195.1	79 **Au** Gold 197.0	80 **Hg** Mercury 200.6	81 **Tl** Thallium 204.4	82 **Pb** Lead 207.2	83 **Bi** Bismuth 209.0	84 **Po** Polonium (209)	85 **At** Astatine (210)	86 **Rn** Radon (222)
7	87 **Fr** Francium (223)	88 **Ra** Radium (226)	89 **Ac*** Actinium (227)	104 **Rf** Rutherfordium (261)	105 **Db** Dubnium (262)	106 **Sg** Seaborgium (266)	107 **Bh** Bohrium (264)	108 **Hs** Hassium (277)	109 **Mt** Meitnerium (268)	110 **Ds** Darmstadtium (271)	111 **Rg** Roentgenium (272)	112 **Cn** Copernicium 112	**Fl** flerovium 114			**Lv** livermorium 116		

Lanthanides / Actinides

58 **Ce** Cerium 140.1	59 **Pr** Praseodymium 140.9	60 **Nd** Neodymium 144.2	61 **Pm** Promethium 144.9	62 **Sm** Samarium 150.4	63 **Eu** Europium 152.0	64 **Gd** Gadolinium 157.2	65 **Tb** Terbium 158.9	66 **Dy** Dysprosium 162.5	67 **Ho** Holium 164.9	68 **Er** Erbium 167.3	69 **Tm** Thulium 168.9	70 **Yb** Ytterbium 173.0	71 **Lu** Lutetium 175.0
90 **Th** Thorium 232.0	91 **Pa** Protactinium (231)	92 **U** Uranium 238.1	93 **Np** Neptunium (237)	94 **Pu** Plutonium (242)	95 **Am** Americium (243)	96 **Cm** Curium (247)	97 **Bk** Berkelium (245)	98 **Cf** Californium (251)	99 **Es** Einsteinium (254)	100 **Fm** Fermium (253)	101 **Md** Mendelevium (256)	102 **No** Nobelium (254)	103 **Lr** Lawrencium (257)

Data booklet

Physical constants

Avogadro constant (Na)	$6.02 \times 10^{23} \, mol^{-1}$
Gas constant (R)	$8.134 \, J \, mol^{-1} \, K^{-1}$
Molar volume of ideal gas: at r.t.p.	$24.0 \, dm^3 \, mol^{-1}$
Specific heat capacity of water (c)	$4.18 \, J \, g^{-1} \, K^{-1}$
Ionic product of water (K_w)	$1.00 \times 10^{-14} \, mol^2 \, dm^{-6}$ at 298 K
1 tonne = $10^6 \, g$	

Arrhenius equation $k = Ae^{-Ea/RT}$ or $\ln k = -Ea/RT + \ln A$

Infrared spectroscopy

Characteristic infrared absorptions in organic molecules

Bond	Wavenumber range/cm^{-1}
C—C Alkanes, alkyl chains	750–1100
C—X Haloalkanes (X—Cl, Br, I)	500–800
C—F Fluoroalkanes	1000–1300
N—H Amine, amide	3300–3500
O—H Alcohols and phenols Carboxylic acids	3200–3600 2500–3300 (broad)
C=C Alkenes Arene	1620–1680 Several peaks in range of 1450–1650 (variable)
C—O Alcohols, esters, carboxylic acids	1000–1350
C=O Aldehydes, ketones, carboxylic acids, acyl chlorides, esters, amides, acid anhydrides	1630-1820
Triple bond stretching vibrations CN CC	 2220-2260 2260–2100

^{1}H nuclear magnetic resonance chemical shifts relative to tetramethylsilane (TMS)

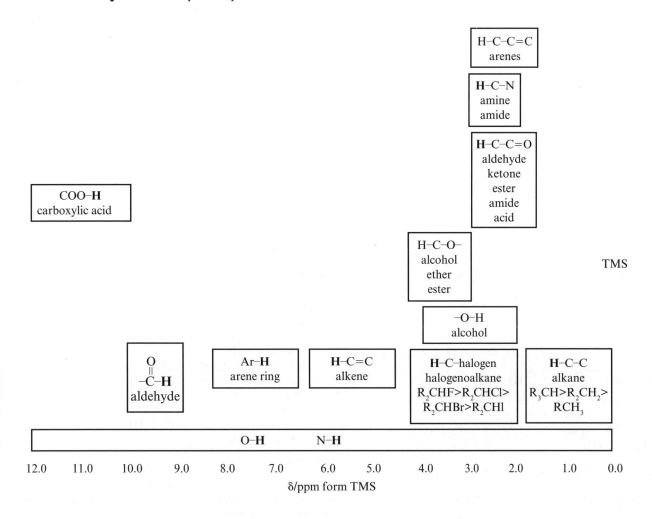

δ/ppm form TMS

^{13}C nuclear magnetic resonance chemical shifts relative to tetramethylsilane (TMS)

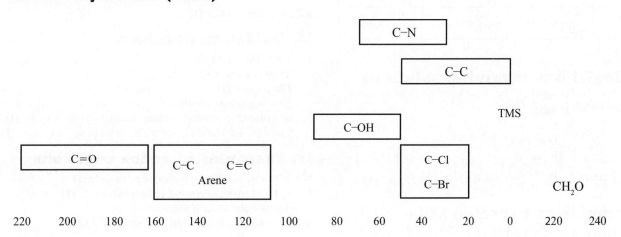

Chemical shifts are variable and can vary depending on the solvent, concentration and substituents. As a result, shifts may be outside the ranges indicated above.

OH and NH chemical shifts are very variable and are often broad. Signals are not usually seen as split peaks.

Note that CH bonded to 'shifting groups' on either side, e.g. O–CH$_2$–C=O, may be shifted more than indicated above.

Answers

1. Atomic structure and isotopes

(a) ^{64}Cu **(1)**
(b) $^{23}Na^+$ **(1)**
(c) $^{35}Cl^-$ **(1)**
(d) $^{16}O^+$ **(1)**

2. Relative masses

392.0 **(1)**

3. Using mass spectroscopy

$69.7 = \dfrac{[69x + 71(100 - x)]}{100}$ **(1)**

$x = 65$ **(1)**
65% Ga-69, 35% Ga-71 **(1)**

4. Writing formulae and equations

$2KOH + H_2SO_4 \rightarrow K_2SO_4 + 2H_2O$
correct reactants and products **(1)**
balancing **(1)**

5. Amount of substance – the mole

(a) Molar mass $= \dfrac{\text{mass}}{\text{moles}} = \dfrac{27.4}{0.28}$ **(1)**

$97.9\,g\,mol^{-1}$ **(1)**

(b) Molar mass $= \dfrac{\text{mass}}{\text{moles}} = \dfrac{49.6}{0.62}$ **(1)**

$80.0\,g\,mol^{-1}$ **(1)**

6. Calculating quantities in reactions

Amount of calcium nitrate $= \dfrac{4.2}{164.1} = 0.0256\,mol$ **(1)**

So amount of all gases $= 0.0256 \times 2.5 = 0.103\,mol$ **(1)**
So volume $= 0.064 \times 24000 = 1536\,cm^3$ **(1)**

7. Types of formulae

	Hydrogen	Sulfur	Oxygen
Amount **(1)**	$\dfrac{0.5}{1} = 0.5$	$\dfrac{8.18}{32} = 0.256$	$\dfrac{16.3}{16} = 1.019$
Ratio **(1)**	$\dfrac{0.5}{0.256} = 1.95$	$\dfrac{0.256}{0.256} = 1$	$\dfrac{1.019}{0.256} = 3.98$

H_2SO_4 **(1)**

8. Calculations involving solutions

Amount of $CO_2 = \left[\dfrac{50}{24000}\right]$ **(1)**

Amount of $HNO_3 = \left[\dfrac{50}{24000}\right] \times 2$ **(1)**

Volume of $HNO_3 = \dfrac{\left[\left[\dfrac{50}{24000}\right] \times 2\right]}{0.20} = 0.021\,dm^3$ or $21\,cm^3$ **(1)**

9. Formulae of hydrated salts

Mass of water $= 4.74 - 3.58 = 1.16\,g$

So amount of water $= \dfrac{1.16}{18} = 0.064\,mol$ **(1)**

So amount of $XCl_2 = 0.032\,mol$ **(1)**

Relative formula mass $= \dfrac{3.58}{0.032} = 111.1$ **(1)**

$111.1 - 71 = 40.1$ so $X = Ca$ **(1)**

10. Percentage yield and atom economy

1 Amount of $C_2H_4 = \dfrac{10}{28}\,mol$ = expected yield of C_2H_5OH **(1)**

Actual yield of $C_2H_5OH = \dfrac{13.7}{46}\,mol$ **(1)**

So % yield $= \left[\dfrac{\left(\dfrac{13.7}{46}\right)}{\left(\dfrac{10}{28}\right)}\right] \times 100 = 83.4\%$ **(1)**

2 Atom economy $\dfrac{92}{180}$ **(1)**

51.1 % **(1)**

11. Neutralisation reactions

1 Lone pair from oxygen forms bond to hydrogen ion. **(1)**
The bond is a dative covalent bond. **(1)**

2 (a) $Na_2CO_3 + 2HCl \rightarrow 2NaCl + H_2O + CO_2$
   correct reactants and products **(1)**
   balancing **(1)**
  (b) $2NH_3 + H_2SO_4 \rightarrow (NH_4)_2SO_4$
   correct reactants and products **(1)**
   balancing **(1)**
  (c) $CuO + 2HNO_3 \rightarrow Cu(NO_3)_2 + H_2O$
   correct reactants and products **(1)**
   balancing **(1)**

12. Acid–base titrations

Titre is proportional to the amount, in mol, of reactant in the flask. **(1)** Rinsing changes the concentration but **not** the amount of reactant in the flask. **(1)**

13. Calculations based on titration data

Amount of KOH $= \left(\dfrac{17.8}{1000}\right) \times 0.250\ (mol)$ **(1)**

Amount of HNO_3 = amount of KOH (1:1 reaction) **(1)**

so $[HNO_3] = \left(\dfrac{17.8}{1000}\right) \times 0.250 \times 50$ **(1)**

$0.223\,mol\,dm^{-3}$ (3 s.f.) **(1)**

14. Oxidation numbers

1 K +1, Mn +7, O −2
 All three correct **(2)**
 Two correct **(1)**
 Zero or one correct **(0)**

2 Br oxidised (oxidation number increases from −1 to 0). **(1)**
 S reduced (oxidation number decreases from +6 to +4). **(1)**

15. Examples of redox reactions

Mn: +7 to +2 (oxidation number decreases by 5)
Fe: +2 to +3 (oxidation number increases by 1) **(1)**
So 5 Fe particles to each Mn particle
$MnO_4^- + 8H^+ + 5Fe^{2+} \rightarrow Mn^{2+} + 5Fe^{3+} + 4H_2O$ **(1)**

17. Electron shells and orbitals

If n = principal quantum number of the energy level, then maximum number of electrons it can hold $= 2n^2$. **(1)**

18. Electron configurations – filling the orbitals

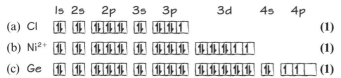

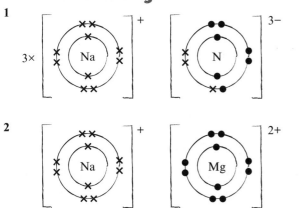

(a) Cl ... **(1)**

(b) Ni^{2+} ... **(1)**

(c) Ge ... **(1)**

19. Ionic bonding

1

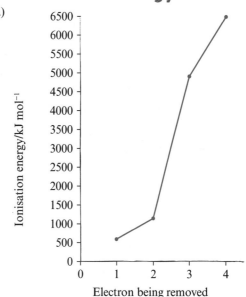

2

Sodium and magnesium ions are isoelectronic as they both have the same number of electrons. **(1)**

20. Covalent bonds

1 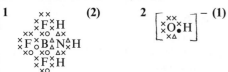 **(2)**

2 $\left[\begin{smallmatrix} \times\times \\ \times O \cdot H \\ \times\triangle \end{smallmatrix}\right]^{-}$ **(1)**

21. Shapes of molecules

1 $\begin{smallmatrix} \times & \times & \bullet\bullet & \times & \times \\ \times O & \vdots & S & \vdots & O \times \\ \times & \bullet & & \bullet & \times \end{smallmatrix}$ **(1)**

There are three regions of electron density so the shape is angular (bent). **(1)** OSO angle is about 117.5° due to additional repulsion of lone pair on S.

2 Tetrahedral **(1)**
109.5° **(1)**

22. More shapes of molecules and ions

Tetrahedral **(1)** 109.5° **(1)**

23. Electronegativity and bond polarity

Non-polar as the bonds are arranged symmetrically. **(1)**
So bond dipoles cancel out. **(1)**

24. Van der Waals' forces

Methane, ethane, propane, butane. **(1)**
Methane has the lowest London forces as it has the lowest number of electrons per molecule; hence its boiling point is the lowest. **(1)**

25. Hydrogen bonding and the properties of water

HF has weaker London forces **(1)** as it has fewer electrons per molecule than HCl **(1)**; however, unlike HCl it can form strong hydrogen bonds so its boiling point is higher than might be expected. **(1)**

26. Properties of simple molecules

None of the electrons in the structure are free to move. **(1)**

28. The Periodic Table

1 The highest energy electron is in a p orbital. **(1)**
2 Periodicity is the gradual change in a property across a period **(1)** that is repeated across each period **(1)**. For example, the atomic radius of atoms across Period 3 gradually falls. **(1)**

29. Ionisation energy

(a)

(b) There are 2 outer electrons **(1)** because there is a sudden jump at the third ionisation energy. **(1)**
(c) For example, magnesium **(1)** because it is in Group 2 with 2 outer electrons. **(1)**

30. Ionisation energy across Periods 2 and 3

1 A (Be) **(1)**
2 A (Li) **(1)**

31. Structures of the elements

(a) $Cl_2(s) \rightarrow Cl_2(l)$ **(1)**
(b) Si–Si bonds are strong **(1)** and need a lot of energy to break, hence a high temperature, when the liquid forms **(1)** but only weak intermolecular forces are broken when chlorine melts. **(1)**

32. Properties of the elements

1 **(a)** Aluminium has a metallic lattice with a strong electrostatic attraction between positive ions and delocalised electrons, so a high boiling point. **(1)**
Silicon has a giant covalent lattice with strong covalent bonds throughout, so a very high boiling point. **(1)**
Chlorine has strong bonds in its molecules **(1)** but only weak intermolecular forces that are broken when boiling, so a low boiling point. **(1)**
(b) Aluminium's electrons can move **(1)** so aluminium is a good conductor. **(1)**
Silicon is a metalloid **(1)** and a limited number of electrons can move through its structure so it has a low conductivity. **(1)**
Chlorine has no free electrons or ions so it is a non-conductor. **(1)**

33. Group 2 elements

1 Bright, white light **(1)** and white ash. **(1)**
2 Magnesium atoms lose electrons and so are oxidised **(1)**, oxygen atoms gain electrons and so are reduced. **(1)**

34. Group 2 compounds and their uses

(a) $BaO(s) + H_2O(l) \rightarrow Ba(OH)_2(aq)$ **(1)**

(b) 13–14 **(1)**

(c) pH is lower **(1)** because calcium hydroxide is less soluble **(1)** so there are fewer hydroxide ions present. **(1)**

35. The halogens and their uses

Chlorine kills bacteria in the water **(1)** because it is toxic **(1)**. However, it may form chlorinated hydrocarbons **(1)**, which may be carcinogenic. **(1)**

36. Reactivity of the halogens

Add the salt solutions to separate test tubes, then add bromine solution and shake. **(1)** In NaCl(aq) the mixture turns yellow/orange because there is no reaction. **(1)** In NaI(aq) the mixture turns brown **(1)** because iodine is formed. **(1)**

$2NaI(aq) + Br_2(aq) \rightarrow 2NaBr(aq) + I_2(aq)$ **(1)**

Bromine is more reactive than iodine (as it displaces iodine) but less reactive than chlorine (as it does not displace chlorine). **(1)**

37. Tests for ions

1 $2AgNO_3(aq) + Na_2CO_3(aq) \rightarrow 2NaNO_3(aq) + Ag_2CO_3(s)$
   formulae **(1)**
   state symbols and balancing **(1)**

2 White precipitate

3 (Pale) yellow precipitate **(1)** due to yellow AgI(s) **(1)**, mixed with white $Ag_2CO_3(s)$ **(1)**; after acid is added, precipitate becomes yellow(er) due to silver carbonate reacting. **(1)**

39. Enthalpy profile diagrams

1 $C_2H_5OH(l) + 3O_2(g) \rightarrow 2CO_2(g) + 3H_2O(l)$
   formulae **(1)**
   balancing **(1)**

2
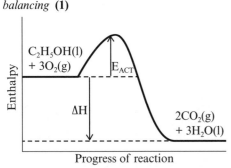

   shape **(1)** ΔH *label* **(1)** E_{ACT} *label* **(1)** *axes labelled* **(1)**

3 Exothermic **(1)** because enthalpy of products is below that of reactants. **(1)**

4 The molecules do not have the necessary activation energy **(1)** until supplied with heat energy from match. **(1)**

40. Enthalpy change of reaction

Place a polystyrene cup into a beaker. **(1)** Add some distilled water measured with a measuring cylinder. **(1)** Measure the temperature of the water. **(1)** Weigh out some ammonium chloride. **(1)** Add to the water and stir vigorously. **(1)** Measure the temperature each minute for 10 minutes. **(1)**

41. Calculating enthalpy changes

$q = mc\Delta T = 100 \times 4.18 \times 55.8$ **(1)**
$\quad = 23324.4 \, J$ **(1)**

$M_r(C_6H_{12}O_6) = 180$; moles of glucose $= \dfrac{1.5}{180} = 0.008333 \, mol$ **(1)**

$q = \dfrac{23324.4}{0.008333} = 2798.928 \, J\,mol^{-1}$ **(1)**

$\Delta H = -2880 \, kJ\,mol^{-1}$ **(1)**

42. Enthalpy change of neutralisation

(a) $q = mc\Delta T = (25 + 23.4) \times 4.18 \times 12.8 = 2589.5936 \, J$ **(1)**
$\quad = 2.59 \, kJ$ **(1)**

(b) moles HCl $= \left(\dfrac{25}{1000}\right) \times 2 = 0.050 \, mol$ **(1)**

$\Delta H = -\dfrac{2.59}{0.05}$ **(1)**

$\quad = -51.8 \, kJ\,mol^{-1}$ **(1)**

43. Hess' law

$\Delta H^{\ominus} = +180 - 112$ **(1)**
$\quad = +68 \, kJ\,mol^{-1}$ **(1)**

44. Enthalpy change of formation

$\Delta_r H = \Sigma \, \Delta_f H \, (products) - \Sigma \, \Delta_f H \, (reactants)$

$\Sigma \, \Delta_f H \, (products) = -718$ **(1)**

$\Sigma \, \Delta_f H \, (reactants) = 3 \times -217 + 0 = -651$ **(1)**

$\Delta_r H = -718 - (-651) = -67 \, kJ\,mol^{-1}$ **(1)**

45. Enthalpy change of combustion

(a) The more carbons in the alkane, the more exothermic the enthalpy of combustion. **(1)**

(b) $\Delta_r H = \Sigma \, \Delta_c H \, (reactants) - \Sigma \, \Delta_c H \, (products)$
$\Sigma \Delta_c H \, (reactants) = -2220$ **(1)**
$\Sigma \Delta_c H \, (products) = -1410 - 890 = -2300$ **(1)**
$\Delta_r H = -2220 - (-2300) = +80 \, kJ\,mol^{-1}$ **(1)**

46. Bond enthalpies

(a) $CH_2=CH_2 + Br_2 \rightarrow CH_2BrCH_2Br$

(b) $\Delta_r H = (4 \times 413 + 614) + 193 - (4 \times 413 + 347 + 2 \times 276)$ **(1)**
$\quad = 2459 - 2551 = -92 \, kJ\,mol^{-1}$ **(1)**

47. Collision theory

Increase concentration of acid **(1)**, more frequent collisions of particles in solution with magnesium. **(1)**

Increase temperature of acid **(1)**, greater proportion of collisions between particles will have at least activation energy and be successful. **(1)**

Use magnesium powder **(1)**, larger surface area means more frequent collisions between particles in solution and surface. **(1)**

48. Measuring reaction rates

(a)
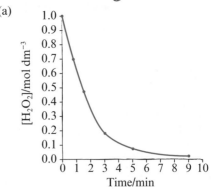

   axes **(1)** *points* **(1)** *curve* **(1)**

(b)
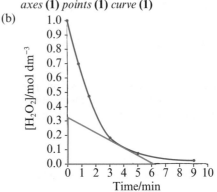

rate, from triangle on graph = 0.32 / 6 = 0.05 mol dm⁻³ min⁻¹ *tangent* (1) *figures* (1) *answer and units* (1)
(c) Collect the oxygen gas evolved (1) in a gas syringe. (1)

49. The Boltzmann distribution
(a) *y*-axis: number of particles (1), *x*-axis: energy (1)
(b) **(2)**

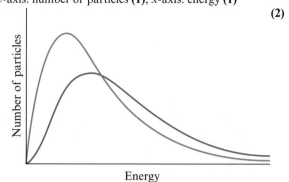

At the higher temperature, a greater area under the curve at $\geq E_a$ so greater proportion of the particles have $\geq E_a$ (1) so a greater proportion of the collisions will be successful. (1) Peak lower (1), peak to right (1).

50. Catalysts
1 $2I^-(aq) + S_2O_8^{2-}(aq) \rightarrow I_2(aq) + 2SO_4^{2-}(aq)$ (1)
2 In the uncatalysed reaction negative ions must collide. (1) In the catalysed reaction, in each step negative and positive ions must collide. (1) As like charges repel and unlike charges attract, the collision of oppositely charged ions requires a lower activation energy. (1)
3 Homogeneous because the catalyst and the reactants are all in aqueous solution and would mix. (1)

51. Dynamic equilibrium
(a) All three gases – $H_2(g), I_2(g)$ and $HI(g)$ – are found. (1)
(b) The forward and backward reactions are still occurring. (1)

52. Le Chatelier's principle
B (1)

53. The equilibrium constant
D (1)

55. Key terms in organic chemistry
1 $Cl_2 \rightarrow 2Cl$ (1)
2 Heterolytic bond fission, as Cl has a greater electronegativity so this suggests it will pull bond pair towards itself, making the bond polar. (1)
   This is more likely to lead to the formation of Cl^- and H^+ ions if the bond breaks. (1)
3 Aliphatic (1), carboxylic acid (1)

56. Naming hydrocarbons
(a) **(1)**

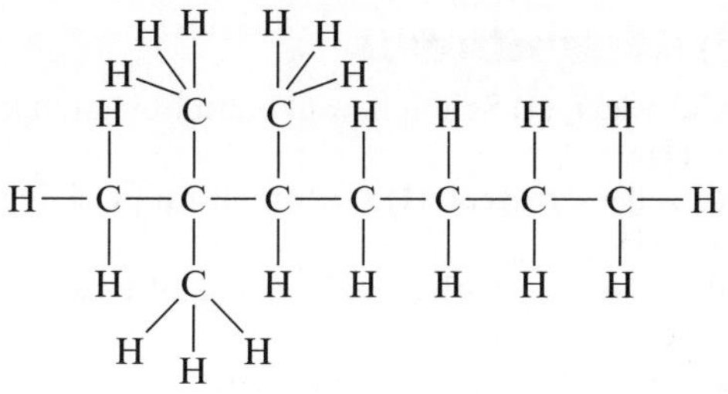

(b) **(1)**

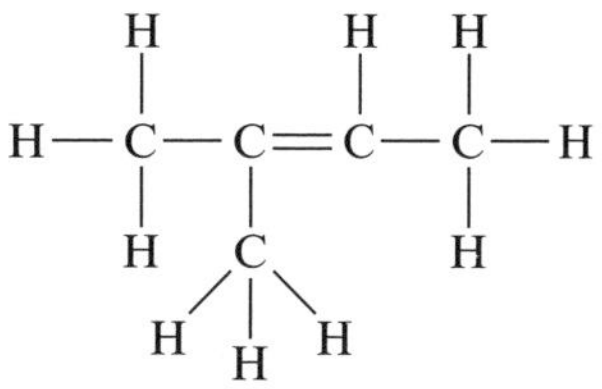

(c) **(1)**

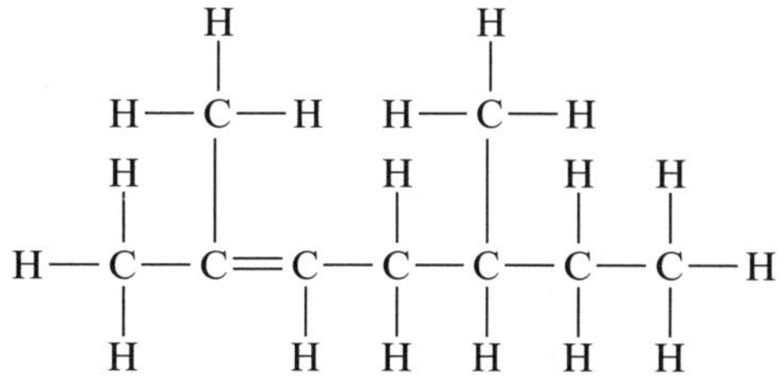

57. Naming compounds with functional groups
(a) **(1)** 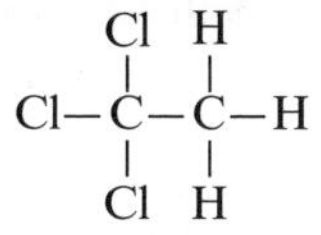
(b) **(1)** 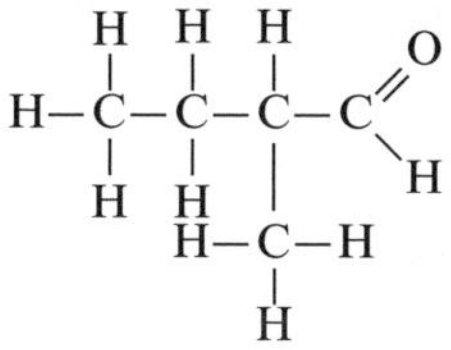
(c) **(1)** 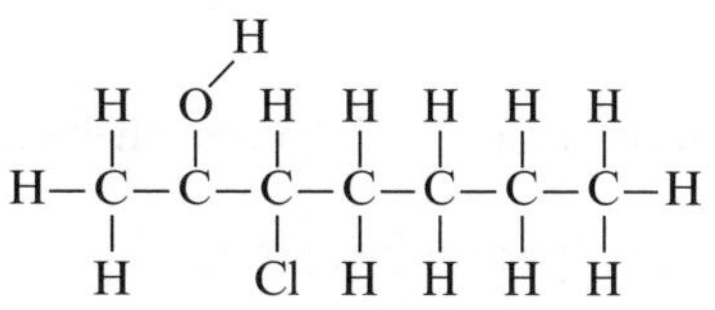
(d) **(1)**

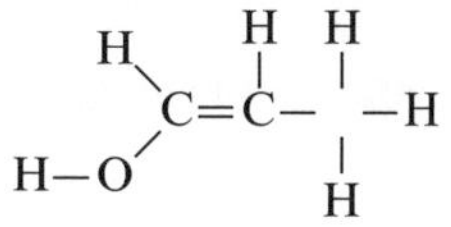

58. Different types of formulae
(a) **(1)**

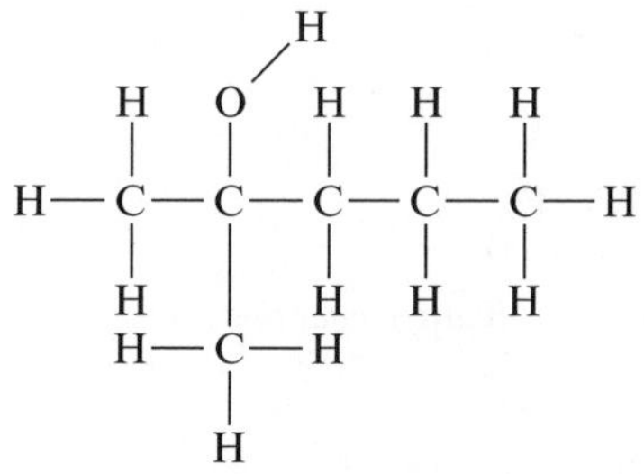

$CH_3C(OH)(CH_3)CH_2CH_2CH_3$ **(1)**

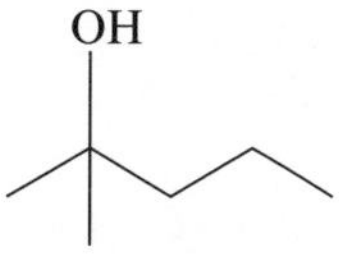

(1)

(b) **(1)**

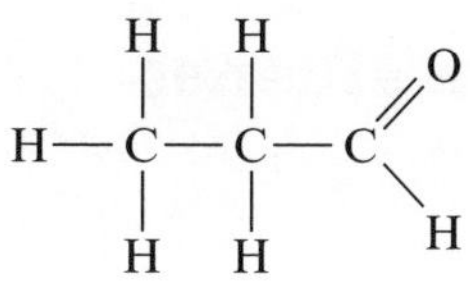

CH_3CH_2CHO **(1)**

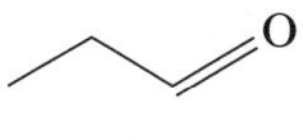

(1)

(c) 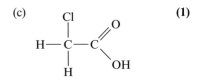 **(1)**

CH$_2$ClCO$_2$H **(1)**

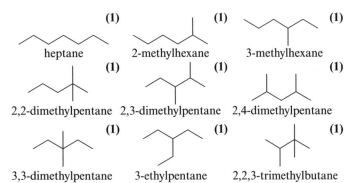 **(1)**

59. Structural isomers

heptane **(1)**   2-methylhexane **(1)**   3-methylhexane **(1)**

2,2-dimethylpentane **(1)**   2,3-dimethylpentane **(1)**   2,4-dimethylpentane **(1)**

3,3-dimethylpentane **(1)**   3-ethylpentane **(1)**   2,2,3-trimethylbutane **(1)**

60. Properties and reactivity of alkanes

1   2-methylhexane, as it has less branching, so more points of contact between molecules **(1)**, hence greater London forces. **(1)**
2   No, as C–H bonds are strong **(1)** and non-polar. **(1)**

61. Reactions of alkanes

1   Initiation
Cl$_2$ → 2Cl• **(1)**
Propagation
CH$_4$ + Cl• → CH$_3$ + HCl **(1)**
CH$_3$• + Cl$_2$ → CH$_3$Cl + Cl• **(1)**
Termination *(any correct termination equation)* **(1)**
2Cl• → Cl$_2$
CH$_3$• + Cl• → CH$_3$Cl
CH$_3$• + CH$_3$• → CH$_3$CH$_3$
2   C$_3$H$_8$ + 2O$_2$ → 3C + 4H$_2$O
correct reactants and products **(1)**
balancing **(1)**

62. Bonding in alkenes

There are three areas of electron density around carbon atom **(1)**, so shape will be trigonal planar. **(1)**

63. Stereoisomerism in alkenes

1

2   *Z*-2-chlorobut-2-ene **(1)**
3   No, because the two groups attached to each of the C=C carbons **(1)** are not the same on both carbons. **(1)**

64. Addition reactions of alkenes

Pair of electrons moving in HBr should move from the bond **not** the hydrogen atom **(1)**; pair of electrons should move from the bromide ion to the carbocation **(1)**.

65. Formation and disposal of polymers

Disadvantages: non-biodegradable **(1)**; made from a finite resource (oil) **(1)**.

Advantages – any two from: low density compared to strength; can be recycled; properties can be designed for specific uses **(2)**.

Chemists can develop polymers made from sustainable raw materials such as plant starch **(1)**.

Chemists can help to dispose of used polymers effectively, e.g. efficient energy recovery by incineration. **(1)**

67. The properties of alcohols

As the ethanol burns, some of the heat produced is absorbed by the water, causing it to turn to water vapour. **(1)**

This keeps the temperature of the note below that required for it to catch fire. **(1)**

68. Combustion and oxidation of alcohols

1   C$_2$H$_5$OH+ 2O$_2$ → 2CO + 3H$_2$O **(2)**

69. More reactions of alcohols

1   1-chloropentane **(1)**
2   Either H$_3$PO$_4$ or H$_2$SO$_4$ **(1)** and heat **(1)**
3   (a) CH$_3$OH + [O] → HCHO
    (b) CH$_3$CHOHCH$_2$CH$_3$ + [O] → CH$_3$COCH$_2$CH$_3$ + H$_2$O **(1)**
    (c) CH$_2$OHCH$_2$OH + 4[O] → CO$_2$HCO$_2$H + 2H$_2$O **(1)**

70. Nucleophilic substitution reactions of haloalkanes

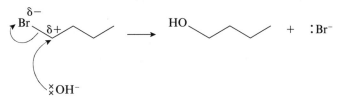

Dipoles shown correctly on C–Br bond, arrow from bond to Br. **(1)** *Arrow from lone pair on hydroxide ion to C in C–Br bond.* **(1)** *Correct products.* **(1)**

71. Preparing a liquid haloalkane

Washing rinses away any impurities that are soluble in the washing solvent. **(1)**

Drying ensures any solvent (e.g. used for washing) evaporates off / is absorbed. **(1)**

Distillation removes any miscible liquid impurities. **(1)**

72. Haloalkanes in the environment

Any HFC (hydrofluorocarbon) **(1)** These do not contain weak C–Cl bonds, so no chlorine free radicals form in the upper atmosphere. **(1)**

73. Organic synthesis

(a) Add bromine water. **(1)** Solution would turn from orange to colourless. **(1)**
(b) Warm with Tollens' reagent. **(1)** A silver mirror precipitate would form. **(1)**
(c) Warm with AgNO$_3$(aq) with a small amount of ethanol. **(1)**
    Then any one of these: **(1)**
    Chloroalkanes would give a white precipitate.
    Bromoalkanes would give a cream precipitate.
    Iodoalkanes would give a yellow precipitate.

75. Infrared spectroscopy

Key absorptions at 1630 to $1820\,cm^{-1}$ ($C=O$ group) (1) and $3200–3600\,cm^{-1}$ (O–H group in an alcohol). (1)
The peak due to the OH group in butanoic acid would be broader and would be seen at a lower wavenumber range, around $2500–3300\,cm^{-1}$. (1)

76. Uses of infrared spectroscopy

Toxic gases, such as CO (1). Gases that may cause photochemical smog, such as NO (1). Gases that contribute to the Greenhouse Effect, such as CO_2 (1).

77. Mass spectrometry

Peak A
$C_2H_3^+$ (1); $C_7H_{14}^+ \rightarrow C_2H_3^+ + C_5H_{11}$ (1)
Peak B
$C_3H_5^+$ (1); $C_7H_{14}^+ \rightarrow C_3H_5^+ + C_4H_9$ (1)
Peak C
$C_7H_{14}^+$ (1); $C_7H_{14} \rightarrow C_7H_{14}^+ + e^-$ (1)

78. Concentration–time graphs (zero order reactants)

Zero order means that a change in the concentration of that reactant does not affect the rate. (1)
Carry out the reaction with a large excess of all the reactants except the one thought to be zero order, measuring the concentration of this reactant. (1)
Plot the concentration of this reactant against time; if it is zero order, the graph will be a straight line. (1)

79. Concentration–time graphs (first order reactants)

Mix a measured amount of gentian violet, in large excess, with sodium hydroxide solution. (1)
At regular intervals, use a pipette to take samples of the reaction mixture. (1)
Quench then titrate each sample to find the concentration of the sodium hydroxide. (1)
Plot a graph of concentration against time; a constant half-life will show that the reaction is first order with respect to hydroxide ions. (1)

80. Rate equation and rate constant

Rate = $15[A][B]^2$ (1)
$= 15 \times 0.05 \times (0.08)^2 = 0.0048\,mol\,dm^{-3}\,s^{-1}$ (1)

81. Finding the order

NO: order = 2 (1) H_2: order = 0 (1) rate = $k[NO]^2$ (1)

82. The rate-determining step

1 HBr: order = 1, (1) O_2: order = 1 (1), overall order = 2 (1)
2 $4HBr + O_2 \rightarrow 2Br_2 + 2H_2O$ (1)

83. The Arrhenius equation

C (1)

85. Finding the equilibrium constant

$$K_c = \frac{\left(\dfrac{2.4}{v}\right)^2}{\left(\dfrac{2.4}{v}\right)\left(\dfrac{1.8}{v}\right)}$$ (1)

$= 1.33$ (1)

86. Calculating the equilibrium constant, K_c

(a) $K_c = \dfrac{[SO_3]^2}{[SO_2]^2[O_2]}$ (1) units = $dm^3\,mol^{-1}$ (1)

(b) $K_c = \dfrac{[NO_2]^2}{[N_2O_4]}$ (1) units = $mol\,dm^{-3}$ (1)

(c) $K_c = \dfrac{[Zn^{2+}]}{[Cu^{2+}]}$ (1) no units (1)

87. Calculating K_p

(a) $SO_2 = \dfrac{0.75}{8.43} = 0.089$

$O_2 = \dfrac{0.18}{8.43} = 0.021$

$SO_3 = \dfrac{7.5}{8.43} = 0.89$ (1)

(b) $SO_2 = 107\,kPa$
$O_2 = 25.6\,kPa$
$SO_3 = 1068\,kPa$ (1)

(c) $K_P = \dfrac{P(SO_3)^2}{P(SO_2)^2 P(O_2)}$ (1)

$= \dfrac{(1068)^2}{(107)^2(25)}$ (1)

$= 3.99\,kPa^{-1}$ (1)

88. The equilibrium constant under different conditions

A (1)

89. Brønsted–Lowry acids and bases

Pair 1: acid H_2SO_4; conjugate base HSO_4^- (1)
Pair 2: base HNO_3 (1); conjugate acid H_3O^+ (1)

90. pH

1 (a) 1.70 (1)
(b) Amount HCl = 0.01 mol, amount NaOH = 0.006 mol (1)
$[H^+] = \dfrac{(0.01 - 0.006)}{0.110} = 0.0364$ (1)

pH = $-\log(0.0364) = 1.44$ (1)
2 $10^{-2.4} = 4.0 \times 10^{-3}\,mol\,dm^{-3}$ (1)

91. The ionic product of water

1 (a) $-\log\left(\dfrac{1 \times 10^{-14}}{0.02}\right) = 12.3$ (1)

(b) $-\log\left(\dfrac{1 \times 10^{-14}}{0.3}\right)$ (1)

$= 13.5$ (1)

(c) Amount NaOH = 0.02 mol, amount HCl = 0.006 mol (1)
$[OH^-] = \dfrac{0.014}{0.140} = 0.1$ (1)

pH = $-\log\left(\dfrac{1 \times 10^{-14}}{0.1}\right) = 13$ (1)

2 (a) $[H^+] = \sqrt{1.52 \times 10^{-13}} = 3.90 \times 10^{-7}\,mol\,dm^{-3}$ (1)
pH = $-\log(3.90 \times 10^{-7}) = 6.41$ (1)
(b) At a higher temperature the pH is lower so there are more H^+ ions (1) so the dissociation equilibrium has moved right, so the dissociation is endothermic. (1)
(c) $[H^+] = [OH^-]$ (1)

92. The acid dissociation constant

(a) $[H^+]^2 = 1.4 \times 10^{-3} \times 0.01 = 1.4 \times 10^{-5}\ mol^2\ dm^{-6}$ **(1)**

$[H^+] = \sqrt{(1.4 \times 10^{-5})} = 3.74 \times 10^{-3}\ mol\ dm^{-3}$ **(1)**

$pH = -\log(3.74 \times 10^{-3}) = 2.43$ **(1)**

(b) $[H^+] = 10^{-3}\ mol\ dm^{-3}$ **(1)**

$[\text{chloroethanoic acid}] = \dfrac{[H^+]^2}{K_a}$ **(1)**

$= \dfrac{(10^{-3})^2}{1.4 \times 10^{-3}} = 7.14 \times 10^{-4}\ mol\ dm^{-3}$ **(1)**

93. Approximations made in weak acid pH calculations

1 $[H^+] = 10^{-2.93} = 1.17 \times 10^{-3}\ mol\ dm^{-3}$ **(1)**

$K_a = \dfrac{[H^+]^2}{(0.010)}$ **(1)**

$= 1.38 \times 10^{-4}\ mol\ dm^{-3}$ **(1)**

2 $K_a = 10^{-4.82} = 1.51 \times 10^{-5}\ mol\ dm^{-3}$ **(1)**

$[H^+]^2 = 1.51 \times 10^{-5} \times 0.5 = 7.57 \times 10^{-6}\ mol^2\ dm^{-6}$ **(1)**

$pH = -\log(\sqrt{(7.57 \times 10^{-6})}) = 2.56$ **(1)**

94. Buffers

During exercise lactic acid is produced, increasing $[H^+]$ in blood. **(1)** Blood is buffered with a mixture of carbonic acid and hydrogencarbonate ions. **(1)** The H^+ ions from lactic acid react with hydrogencarbonate ions, minimising the pH change in the blood. **(1)**

95. Buffer calculations

1 Original amount butanoic acid $= \dfrac{100}{1000} \times 0.25 = 0.025\ mol$

amount of NaOH $= \dfrac{50}{1000} \times 0.15 = 0.0075\ mol$ **(1)**

After mixing, $[\text{butanoic acid}] = \dfrac{(0.025 - 0.0075)}{0.150}$
$= 0.1167\ mol\ dm^{-3}$;

After mixing, $[\text{sodium butanoate}] = \dfrac{0.0075}{0.150} = 0.0500\ mol\ dm^{-3}$ **(1)**

$K_a = 1.5 \times 10^{-5} = [H^+]\dfrac{(0.05)}{(0.1167)}$

$[H^+] = 5 \times 10^{-5}\ mol\ dm^{-3}$ **(1)**

$pH = -\log(5 \times 10^{-5}) = 4.46$ **(1)**

96. pH titration curves

Start pH $= -\log(0.1) = 1$ **(1)**

end pH: $[OH^-] = \dfrac{(25 \times 0.1)}{75} = 0.0333$

$pH = -\log\left(\dfrac{1 \times 10^{-14}}{0.0333}\right) = 12.5$ **(1)**

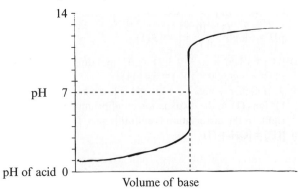

Labelled axes, **(1)** *shape,* **(1)** *vertical region at least pH 4 to 10.* **(1)**

97. Indicators

(a) Strong acid – weak / strong. **(1)**

(b) When the weak acid is neutralised there is a gradual change of pH up to pH 7, **(1)** so the indicator would gradually change colour, changing colour before the equivalence point and not giving a sharp colour change. **(1)**

(c) In a weak acid–weak base titration there is no vertical region on the pH graph, **(1)** so no indicator would give a sharp colour change. **(1)** Instead, a thermometric titration can be carried out, as neutralisation is exothermic, or a pH probe can be used. **(1)**

99. The Born–Haber cycle

(a)

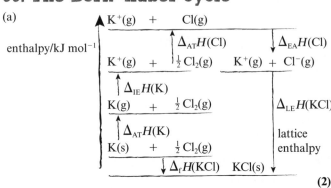

(2)

(b) $349 - 122 - 419 - 77 - 437 = -706$ **(1)** $kJ\ mol^{-1}$ **(1)**

100. Factors affecting lattice enthalpy

Potassium iodide, potassium bromide, potassium chloride, sodium chloride, lithium chloride, lithium oxide. **(1)**

Group I metal ions get larger down the group, so electrostatic attraction to the anion decreases. **(1)**

Halide ions get larger down the group, so electrostatic attraction to the cation decreases. **(1)**

Lithium oxide has the smallest cation and an anion with twice the charge of the halide ions so will have the most exothermic lattice enthalpy. **(1)**

101. The enthalpy change of solution

$\Delta_{sol}H = -\text{lattice enthalpy} + \text{enthalpy change of hydration }(Mg^{2+})$
$+ 2 \times \text{enthalpy change of hydration }(Cl^-)$ **(1)**

$= 2524 + (-1926) + (2 \times -378) = -158\ kJ\ mol^{-1}$ **(1)**

102. Entropy

1 $CH_3CH_2CH_2CH_2OH + 4O_2 \rightarrow 4CO + 5H_2O$ **(1)**

$S\ (\text{products}) = (4 \times 198) + (5 \times 70) = 1142$

$S\ (\text{reactants}) = 228 + (4 \times 205) = 1048$ **(1)**

$\Delta S = 1142 - 1048 = 94\ J\ K^{-1}\ mol^{-1}$ **(1)**

2 Solid wax < liquid wax at 40°C < liquid wax at 60°C < carbon dioxide and water vapour. **(1)**

Gases are more disordered and have higher entropy than liquids, which are more disordered than solids. **(1)** Particles become more disordered as they gain energy, so liquid wax is more disordered, with higher entropy, at 60°C than at 40°C. **(1)**

103. Free energy

(a) $\Delta H^{\ominus} = (-110) - (-635)$ **(1)**
$= 525\ kJ\ mol^{-1}$ **(1)**

(b) $\Delta S\emptyset = (41.4 + 197.6) - (39.7 + 5.7)$ **(1)**
$= 193.6\ J\ K^{-1}\ mol^{-1}$ **(1)**

(c) $\Delta G = \Delta H - T\Delta S = 0$ **(1)**

$T \times \dfrac{193.6}{1000} = 525$ **(1)**

$T = 2712\ K$ **(1)**

104. Redox

$S_2O_8^{2-} + 2Fe^{2+} \rightarrow 2SO_4^{2-} + 2Fe^{3+}$

formulae on correct sides **(1)** balancing **(1)**

105. Redox titrations

Amount $MnO_4^- = \frac{30}{1000} \times 0.0500 = 0.0015$ mol **(1)**

amount $V^{n+} = \frac{25}{1000} \times 0.100 = 0.0025$ mol **(1)**

Each MnO_4^- ion accepts $5e^-$, so moles of electrons gained = moles of electrons lost by $V^{n+} = 5 \times 0.0015 = 0.0075$ mol **(1)**

Moles of electrons per mole of $V^{n+} = \frac{0.0075}{0.0025} = 3$ **(1)**

so vanadium's oxidation state reduces from 5 in VO_2^+ to 2 in V^{n+}; hence $n = 2$ **(1)**

106. Electrochemical cells

(a) $Cl_2 + 2e^- \rightarrow 2Cl^-$; **(1)** $Br_2 + 2e^- \rightarrow 2Br^-$ **(1)**
(b) The chlorine half-cell. **(1)**

107. Measuring and using standard electrode potentials

(a) $2Fe^{3+} + 2I^- \rightarrow 2Fe^{2+} + I_2$ **(1)**
(b) $E^{\ominus} = +0.77 - 0.54 = +0.23$ V **(1)**

108. Predicting feasibility

Fe^{3+} to Fe^{2+} $E^{\ominus} = +0.77$ V, so Fe^{3+} will oxidise V^{2+} to V^{3+} (+1.03 V) and V^{3+} to VO^{2+} (+0.43 V), **(1)** but not VO^{2+} to VO_2^+ (−0.23 V), so final species is VO^{2+} **(1)**

109. Storage and fuel cells

Only emission from vehicle is water, **(1)** but hydrogen is produced by electrolysis **(1)**; carbon dioxide emissions will only be reduced if electricity generated for this is without using fossil fuels. **(1)**

111. The transition elements

Cr $1s^22s^22p^63s^23p^63d^54s^1$ Cr^{3+} $1s^22s^22p^63s^23p^63d^3$ **(1)**
Zn $1s^22s^22p^63s^23p^63d^{10}4s^2$ Zn^{2+} $1s^22s^22p^63s^23p^63d^{10}$ **(1)**
Each atom's highest energy electron is in a 3d orbital, so both are d-block elements. **(1)** Chromium's ion has a part-full 3d sub-shell, but zinc's only ion has a full 3d sub-shell, so only chromium is a transition element. **(1)**

112. Properties of transition elements

Carry out the experiment with a measured volume of acid of known concentration and a measured mass of zinc, collecting the gas in a gas syringe. **(1)** Measure the gas volume in equal intervals of time. **(1)** Repeat under identical conditions, but add a small volume of solution containing copper(II) ions. **(1)** Plot the volume of gas against time for both experiments to show that the one with added copper(II) ions has a larger initial gradient. **(1)**

113. Complex ions

hexaaquairon (II) ion

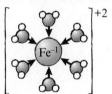

Diagram has correct shape, **(1)** *and co-ordinate bonds shown.* **(1)**
Octahedral, **(1)** *because the six co-ordinate bonds are arranged symmetrically.* **(1)**

114. 4-fold coordination and isomerism

C **(1)**

115. Precipitation reactions

Add ammonia solution, forming a precipitate, then further ammonia, shaking. **(1)** Iron(II) ions form a green precipitate, **(1)** chromium(III) ions form a green precipitate that dissolves in excess ammonia to give a violet solution. **(1)**

116. Ligand substitution reactions

(a) $B \rightarrow C$ $[Cu(H_2O)_6]^{2+} + 2OH^- \rightarrow [Cu(OH)_2(H_2O)_4] + 2H_2O$ **(1)**
$C \rightarrow D$ $[Cu(OH)_2(H_2O)_4] + 4NH_3 \rightarrow [Cu(NH_3)_4(H_2O)_2]^{2+} + 2H_2O + 2OH^-$ **(1)**
$B \rightarrow E$ $[Cu(H_2O)_6]^{2+} + 4Cl^- \rightarrow [CuCl_4]^{2-} + 6H_2O$ **(1)**
(b) B is blue and E is yellow, so when some of B has reacted, the mixture of blue and yellow looks green. **(1)**

117. Redox reactions of transition elements

(a) $MnO_4^- + 8H^+ + 5Fe^{2+} \rightarrow Mn^{2+} + 4H_2O + 5Fe^{3+}$ **(1)**
(b) The MnO_4^- solution is purple and the Mn^{2+} solution is pink. **(1)** The Fe^{2+} solution is green and the Fe^{3+} solution is yellow. **(1)**

119. The bonding in benzene rings

No, as chlorine is non-polar, **(1)** and electron density in benzene ring is not sufficient to polarise the chlorine molecule. **(1)**

120. Reactions of benzene rings

$AlCl_3$ reacts with halogen such as chlorine to form Cl^+:
$AlCl_3 + Cl_2 \rightarrow AlCl_4^- + Cl^+$ **(1)**

But then $AlCl_3$ is reformed by reaction of $AlCl_4^-$ with the H^+ that has been substituted from the ring:
$AlCl_4^- + H^+ \rightarrow AlCl_3 + HCl$ **(1)**

121. Electrophilic substitution reactions

Arrow from ring to SO_3^+, **(1)** formula of ring intermediate, **(1)** arrow back to ring and final products. **(1)**

122. Comparing the reactivity of alkenes and aromatic compounds

Ethene – bromine is decolourised, **(1)** benzene – no visible change. **(1)** Ethene has greater electron density in π bond so polarises Br_2. **(1)**

123. Phenol

Phenol can release a H^+ ion in water. **(1)** The negative charge on the phenoxide ion that is formed, $C_6H_5O^-$, can delocalize into ring, stabilising the ion. **(1)** Ethanol is not likely to release an H^+ ion in water as ethoxide ion is less stable. **(1)**

125. Exam skills 11

The ester group will increase the rate of electrophilic substitution, as it will increase electron density of ring, as it is an electron donating group, **(1)** it will favour substitution at the 2,4 positions in the ring. **(1)**

126. Aldehydes and ketones

As Tollens' reagent is an oxidising agent, **(1)** and there are other compounds that it could also oxidise. **(1)**

127. Nucleophilic addition reactions

(a)
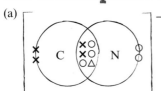

Lone pairs of electrons on both C and N, (1) three pairs of bonding electrons between C and N. (1)

(b)
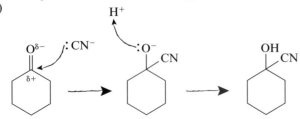

Arrow from lone pair on cyanide ion to partially charged carbon, (1) structure of intermediate, (1) arrow from lone pair to hydrogen ion and final product. (1)

128. Carboxylic acids

(a) $2CH_3CH_2CO_2H + CaCO_3 \rightarrow Ca(CH_3CH_2CO_2)_2 + H_2O + CO_2$
Correct formulae of reactants and products, (1) correct balancing. (1)

(b) $2CH_3CH_2CO_2H + CuO \rightarrow Cu(CH_3CH_2CO_2)_2 + H_2O$
Correct formulae of reactants and products, (1) correct balancing. (1)

129. Esters

(1)

Pentan-1-ol, (1) hexanoic acid. (1)

130. Acyl chlorides

(a) (1)

(b) (1)

131. Amines

(a) Methylamine, (1) 1-bromopropane. (1)
(b) Less basic as lone pair on nitrogen in phenylamine is delocalised into ring, (1) so less likely / available to accept a proton. (1)

132. Amino acids

(1)

Ester functional groups, (1) rest of product. (1)

133. Optical isomers

C (1)

134. Condensation polymers

(1)

136. Carbon–carbon bond formation

$CH_3CH_2OH \rightarrow CH_3CH_2Br$ (1)
$CH_3CH_2Br \rightarrow CH_3CH_2CN$ (1)
$CH_3CH_2CN \rightarrow CH_3CH_2CO_2H$ (1)
$CH_3CH_2CO_2H \rightarrow CH_3CH_2CO_2Cl$ (1)

137. Purifying organic solids

Filter hot solution (1) through warm filter funnel and paper. (1)

138. Predicting the properties and reactions of organic compounds

(a) CH_2BrCH_2Cl/ $CH_3CHBrCl$ (1)
(b) CH_2CHOH (1)
(c) $\{CH_2CHCl\}_n$ (1)

140. Organic synthesis

(a) Keep reaction mixture cool. (1)
(b) $CH_3CHCH_2 \rightarrow CH_3CH_2CH_2OH$ (1) ; H_3PO_4 (aq) and heat (1)
$CH_3CH_2CH_2OH \rightarrow CH_3CH_2CO_2H$ (1) ; $Cr_2O_7^{2-}$ / H^+ and heat (under reflux) (1)
$CH_3CH_2CO_2H \rightarrow CH_3CH_2COCl$ (1) ; $SOCl_2$ (1)
$CH_3CH_2COCl \rightarrow CH_3CH_2CONH_2$ (1) ; NH_3 (1)

141. Thin layer chromatography

$R_f = \dfrac{4.9}{9.3}$ (1) $= 0.53$ (1)

142. Gas chromatography

Oxygen is not inert, (1) so may react with sample. (1)

143. Qualitative tests for functional groups (1)

Add sodium carbonate solution; the carboxylic acid will fizz but the phenol will show no visible change. (1) Add bromine solution; the phenol will form a white precipitate / the bromine decolourises but the carboxylic acid will show no visible change. (1)

144. Qualitative tests for functional groups (2)

Lone pairs of electrons on the oxygen atoms in the carboxylic acid delocalise across the –COOH group. This stabilizes the –COOH group so it does not react with 2,4-dinitrophenylhydrazine.

145. Carbon-13 NMR spectroscopy

(1) (1) (1) (1)

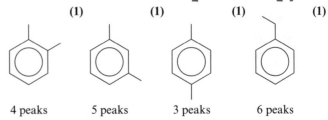

4 peaks   5 peaks   3 peaks   6 peaks

146. Proton NMR spectroscopy

C_6H_5 protons at $\delta = 6.6–8$ ppm, (1) relative area under peak(s) $= 5$ (1)
CHO proton at $\delta = 3.2–4.2$ ppm, (1) relative area under peak $= 1$ (1)

147. Identifying the structure of a compound from a proton (H–1)NMR spectrum

Both spectra will have three peaks (as both structures have three different hydrogen environments). **(1)** Both spectra will have a quartet, a triplet and a singlet. **(1)**

Quartet in spectrum for ethyl ethanoate at $\delta = 3$–4 ppm, for methyl propanaote at $\delta = 2$–3 ppm. **(1)**

Singlet in spectrum for ethyl ethanoate at $\delta = 2$–3 ppm, for methyl propanaote at $\delta = 3$–4 ppm **(1)**

148. Predicting a proton NMR spectrum

Singlet at $\delta = 2$–3 ppm as $COCH_3$ protons have no neighbouring protons on adjacent carbon. **(1)**

Quartet at $\delta = 3.2$–4.2 ppm as $-CH_2O-$ protons have three neighbouring protons on adjacent methyl group. **(1)** Three triplets as three hydrogen environments have adjacent $-CH_2-$ groups. **(1)**

149. Deducing the structure of a compound from a range of data

8 peaks **(1)**

Published by Pearson Education Limited, 80 Strand, London, WC2R 0RL.

Copies of official specifications for all OCR qualifications may be found on the OCR website: www.ocr.org.uk

Text © Pearson Education Limited 2016
Edited by Ros Woodward
Typeset by Kamae Design
Produced by Out of House Publishing
Illustrated by Tech-Set Ltd, Gateshead
Cover illustration © Miriam Sturdee

The rights of Mark Grinsell and David Brentnall to be identified as authors of this work have been asserted by them in accordance with the Copyright, Designs and Patents Act 1988.

First published 2016

19 18 17
10 9 8 7 6 5 4 3

British Library Cataloguing in Publication Data
A catalogue record for this book is available from the British Library

ISBN 978 1 447 98437 5

Printed in UK by Bell and Bain

Acknowledgements
The publisher would like to thank the following organisations for their kind permission to reproduce their photographs:

Getty Images: Stephen Hamilton 34; **Science Photo Library Ltd:** 33, Andrew Lambert Photography: 36, 37.

All other images © Pearson Education

Every effort has been made to trace the copyright holders and we apologise in advance for any unintentional omissions. We would be pleased to insert the appropriate acknowledgement in any subsequent edition of this publication.